JACARANDA

GEOGRAPHY ALIVE 9

VICTORIAN CURRICULUM | SECOND EDITION

JACARANDA

GEOGRAPHY ALIVE 9

VICTORIAN CURRICULUM | SECOND EDITION

JILL PRICE

CATHY BEDSON

DENISE MILES

KINGSLEY HEAD

ALISTAIR PURSER

JANE WILSON

CONTRIBUTING AUTHORS

RACHEL RAMSAY | ALEX ROSSIMEL | ELYSE CHORA

jacaranda
A Wiley Brand

Second edition published 2020 by
John Wiley & Sons Australia, Ltd
42 McDougall Street, Milton, Qld 4064

First edition published 2017

Typeset in 11/14 pt TimesLTStd

© John Wiley & Sons Australia, Ltd 2020

The moral rights of the authors have been asserted.

ISBN: 978-0-7303-7919-5

Front cover image: © David Trood/Getty Images Australia

Illustrated by various artists, diacriTech and Wiley Composition Services

Typeset in India by diacriTech

Printed in Singapore by
Markono Print Media Pte Ltd

A catalogue record for this
book is available from the
National Library of Australia

NATIONAL LIBRARY OF AUSTRALIA

10 9 8 7 6 5 4 3

CONTENTS

9 Trade — a driving force for interconnection

10 Global ICT — connections, disparity and impacts

Fieldwork inquiry: What are the effects of travel in the local community? online only

HOW TO USE
the *Jacaranda Geography Alive* resource suite

The ever-popular *Jacaranda Geography Alive for the Victorian Curriculum* is available as a standalone Geography series or as part of the *Jacaranda Humanities Alive* series, which incorporates Geography, History, Civics and Citizenship, and Economics and Business in a 4-in-1 title. The series is available across a number of digital formats: learnON, eBookPLUS, eGuidePLUS, PDF and iPad app.

Skills development is integrated throughout, and explicitly targeted through SkillBuilders and a dedicated Geographical skills and concepts topic for each year level.

This suite of resources is designed to allow for differentiation, flexible teaching and multiple entry and exit points so teachers can *teach their class their way*.

Features

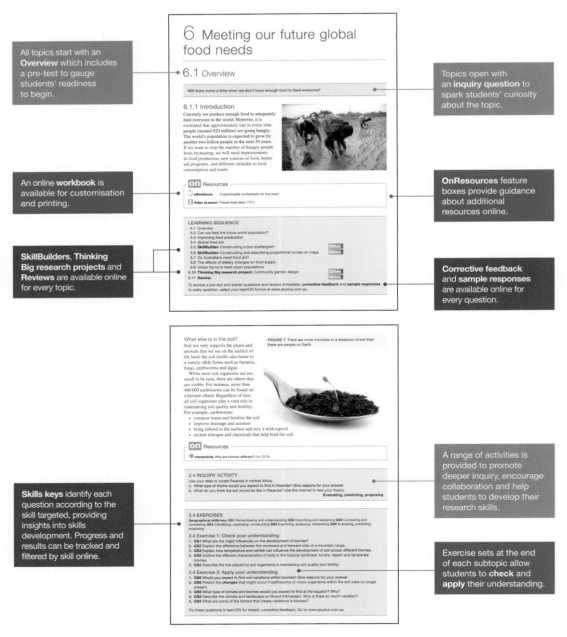

All topics start with an **Overview** which includes a pre-test to gauge students' readiness to begin.

An online **workbook** is available for customisation and printing.

SkillBuilders, Thinking Big research projects and **Reviews** are available online for every topic.

Skills keys identify each question according to the skill targeted, providing insights into skills development. Progress and results can be tracked and filtered by skill online.

Topics open with an **inquiry question** to spark students' curiosity about the topic.

OnResources feature boxes provide guidance about additional resources online.

Corrective feedback and **sample responses** are available online for every question.

A range of activities is provided to promote deeper inquiry, encourage collaboration and help students to develop their research skills.

Exercise sets at the end of each subtopic allow students to **check** and **apply** their understanding.

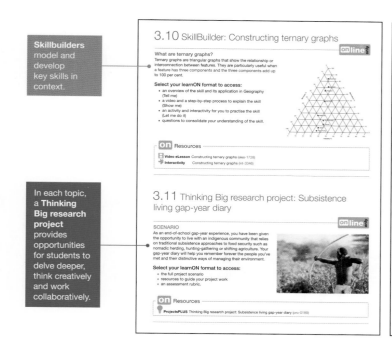

3.10 SkillBuilder: Constructing ternary graphs

What are ternary graphs?
Ternary graphs are triangular graphs that show the relationship or interconnection between features. They are particularly useful when a feature has three components and the three components add up to 100 per cent.

Select your learnON format to access:
- an overview of the skill and its application in Geography (Tell me)
- a video and a step-by-step process to explain the skill (Show me)
- an activity and interactivity for you to practise the skill (Let me do it)
- questions to consolidate your understanding of the skill.

Resources
Video eLesson Constructing ternary graphs (eles-1728)
Interactivity Constructing ternary graphs (int-3346)

3.11 Thinking Big research project: Subsistence living gap-year diary

SCENARIO
As an end-of-school gap-year experience, you have been given the opportunity to live with an indigenous community that relies on traditional subsistence approaches to food security such as nomadic herding, hunting-gathering or shifting agriculture. Your gap-year diary will help you remember forever the people you've met and their distinctive ways of managing their environment.

Select your learnON format to access:
- the full project scenario
- resources to guide your project work
- an assessment rubric.

Resources
ProjectsPLUS Thinking Big research project: Subsistence living gap-year diary (pro-0169)

- improving the food supply chain to remote areas
- establishing and improving community stores in remote areas through initiatives such as the Outback Stores program
- education campaigns about nutrition guidelines and healthy food choices
- grants to support the establishment or improvement of community food initiatives, such as farmers' markets, food cooperatives, food hubs, community gardens and city farms
- food subsidies for fruit and vegetable programs to improve health and nutrition for Indigenous children
- funding for the Stephanie Alexander Kitchen Garden National Program to develop gardens in schools
- improving skills of Indigenous health workers to enable them to have greater influence on food security through addressing health promotion and nutrition at the local level.

FIGURE 7 Papunya Community Store, located 240 kilometres north-west of Alice Springs

Funding has also been made available to various industries that relate to Indigenous peoples' connection to the land. Industry ventures have included park and ranch management to further eco-tourism, and aquaculture. These ventures can also provide food and are a source of revenue for further community development, employment and training opportunities. They can be seen as developing a sense of pride as well as prospects for future generations. They are in keeping with Indigenous peoples' roles as caretakers and traditional custodians of the land.

One such initiative is Fish River Station in the Daly River region of the Northern Territory, home to an array of wildlife, plant life and important traditional food sources for Indigenous peoples. The partnership between government, conservation organisations and the Indigenous Land Corporation aims to preserve this unique environment for future generations, whilst providing employment for Indigenous people and opportunities to reconnect with this culturally significant land.

DISCUSS
What can we learn from the land and resource management practices of Indigenous Australians in relation to food security?   [Intercultural Capability]

4.3.3 What happens when forests are cleared?
FIGURE 4 illustrates some changes that forest clearing in the Amazon can have on the environment.

FIGURE 4 Impacts of clearing the Amazon forest

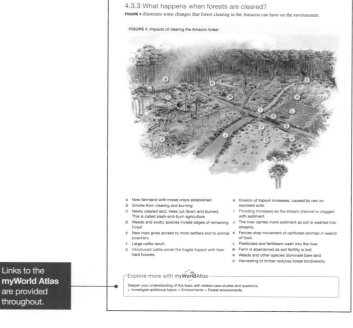

A New farmland with mixed crops established
B Smoke from clearing and burning
C Newly cleared land, trees cut down and burned. This is called slash-and-burn agriculture.
D Weeds and exotic species invade edges of remaining forest
E New road gives access to more settlers and to animal poachers
F Large cattle ranch
G Introduced cattle erode the fragile topsoil with their hard hooves.
H Erosion of topsoil increases, caused by rain on exposed soils.
I Flooding increases as the stream channel is clogged with sediment.
J The river carries more sediment as soil is washed into streams.
K Fences stop movement of rainforest animals in search of food.
L Pesticides and fertilisers wash into the river.
M Farm is abandoned as soil fertility is lost
N Weeds and other species dominate bare land.
O Harvesting of timber reduces forest biodiversity.

Explore more with myWorldAtlas
Deepen your understanding of this topic with related case studies and questions.
- Investigate additional topics > Environments > Forest environments

---

2.12 Review

2.12.1 Key knowledge summary
Use this dot point summary to review the content covered in this topic.
2.12.2 Reflection
Reflect on your learning using the activities and resources provided.

Resources
eWorkbook Reflection (doc-31714)
            Crossword (doc-31715)
Interactivity All the world is a biome crossword (int-7644)

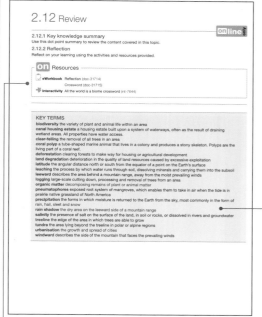

KEY TERMS
biodiversity the variety of plant and animal life within an area
canal housing estate a housing estate built upon a system of waterways, often as the result of draining wetland areas. All properties have water access.
clear-felling the removal of all trees in an area
coral polyp a tube-shaped marine animal that lives in a colony and produces a stony skeleton. Polyps are the living part of a coral reef.
deforestation clearing forests to make way for housing or agricultural development
land degradation deterioration in the quality of land resources caused by excessive exploitation
latitude the angular distance north or south from the equator of a point on the Earth's surface
leaching the process by which water runs through soil, dissolving minerals and carrying them into the subsoil
leeward describes the area behind a mountain range, away from the moist prevailing winds
logging large-scale cutting down, processing and removal of trees from an area
organic matter decomposing remains of plant or animal matter
pneumatophores exposed root system of mangroves, which enables them to take in air when the tide is in
prairie native grassland of North America
precipitation the forms in which moisture is returned to the Earth from the sky, most commonly in the form of rain, hail, sleet and snow
rain shadow the dry area on the leeward side of a mountain range
salinity the presence of salt on the surface of the land, in soil or rocks, or dissolved in rivers and groundwater
treeline the edge of the area in which trees are able to grow
tundra the area lying beyond the treeline in polar or alpine regions
urbanisation the growth and spread of cities
windward describes the side of the mountain that faces the prevailing winds

learnON

Jacaranda Geography Alive learnON is an immersive digital learning platform that enables student and teacher connections, and tracks, monitors and reports progress for immediate insights into student learning and understanding.

It includes:

- a wide variety of embedded videos and interactivities
- questions that can be answered online, with sample responses and immediate, corrective feedback
- additional resources such as activities, an eWorkbook, worksheets, and more
- Thinking Big research projects
- SkillBuilders
- teachON, providing teachers with practical teaching advice, teacher-led videos and lesson plans.

teachON

Conveniently situated within the learnON format, teachON includes practical teaching advice, teacher-led videos and lesson plans, designed to support, save time and provide inspiration for teachers.

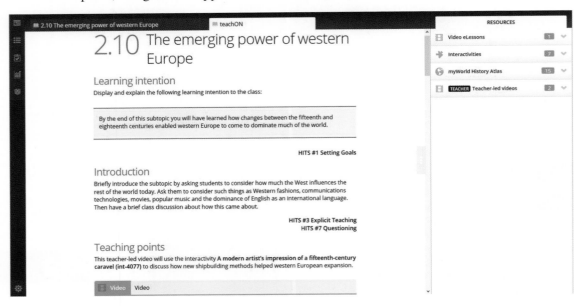

ACKNOWLEDGEMENTS

The authors and publisher would like to thank the following copyright holders, organisations and individuals for their assistance and for permission to reproduce copyright material in this book.

The Victorian Curriculum F–10 content elements are © VCAA, reproduced by permission. The VCAA does not endorse or make any warranties regarding this resource. The Victorian Curriculum F–10 and related content can be accessed directly at the VCAA website. VCAA does not endorse nor verify the accuracy of the information provided and accepts no responsibility for incomplete or inaccurate information. You can find the most up to date version of the Victorian Curriculum at http://victoriancurriculum.vcaa.vic.edu.au.

Images

• AAP Newswire: **166**/Dan Peled • AEGIC **231** (top)/© AEGIC. All rights reserved. • Andrew J. Haysom: **169** (b) • Artizans: **94**/Jake Fuller/artizans.com • AIHW: **181** • Alamy Australia Pty Ltd: **2** (top)/SCPhotos; **8** (bottom)/Dinodia Photos; **52** (top)/AGF Srl; **68**/Nigel Cattlin; **105** (d)/AfriPics.com; **119** (top)/Frans Lanting Studio; **145**/Kim Haughton; **156**/Dinodia Photos; **157**/epa european pressphoto agency b.v.; **161**/ © Hilke Maunder; **189**/© T.M.O.Travel; **209** (top right)/James Osmond; **219**/Elliot Nichol; **259** (top)/ Benedicte Desrus; **249**/Andrew McConnell • Asia Pacific Group: **179** • Australian Bureau of Statistics: **150** (left, right) • Australian Made Campaign Ltd.: **234** (bottom) • Australscope: **108** (top)/Marcos Brindicci/ REUTERS/ Picture Media • Bike Paths & Rail Trails: **174** • City of Ballarat: **183** • Copyright Clearance Center: **185** (bottom)/The complex network of global cargo ship movements, Pablo Kaluza, Andrea Kölzsch, Michael T. Gastner and Bernd Blasius Journal of The Royal Society Interface Volume 7, Issue 48 Published: 19 January 2010 https://doi.org/10.1098/rsif.2009.0495 The complex network of global cargo ship movements • Corbis Australia: **130**/Sally A. Morgan; Ecoscene • Creative Commons: **28**, **34** (bottom), **36**, **62**, **169** (d); **21** (top)/Bureau of Meteorology; **61**/denisbin; **132**/© 2019 Sustainability Victoria; **151**/ Copyright © 2016 International Center for Tropical Agriculture – CIAT; **263**/Source: The Global E-waste Monitor 2017 © Statista; **104** (top)/World Resources Institute; **224** (left)/World Bank national accounts data, and OECD National Accounts data files; **224** (right)/World Bank national accounts data, and OECD National Accounts data files; **209** (bottom)/Source http://www.zoo.org.au/healesville/animals/helmeted-honeyeater The State of Victoria, Department of Environment, Land, Water Planning 2016; **226**, **227** (top), **228** (top)/Based on ABS trade data on DFAT STARS database, ABS catalogue 5368.0.55.003/4 and unpublished ABS data; **227** (bottom)/Department of Education and Training; **241** (top)/© Department of Foreign Affairs and Trade website – www.dfat.gov.au; **253**/International Telecommunication Union, World Telecommunication/ICT Development Report and database; **256** (top, bottom)/Thomas, J, Barraket, J, Wilson, C, Ewing, S, MacDonald, T, Tucker, J & Rennie, E, 2017, *Measuring Australia's Digital Divide: The Australian Digital Inclusion Index 2017*, RMIT University, Melbourne, for Telstra; **264** (top, bottom)/ chinadialogue • David Beaumont: **58** • Department of Agriculture: **234** (middle)/DAFF 2013, *Australian food statistics 2011–12*, Department of Agriculture, Fisheries and Forestry, Canberra. CC BY 3.0 • Department of Environment, Land, Water & Planning : **177** (top) • FAO: **43** (bottom)/Source: Food and Agriculture Organization of the United Nations, 2015, *World agriculture: towards 2015/2030 — Summary report*, Table: Crop yields in developing countries, 1961 to 2030; **90**/*OECD-FAO Agricultural Outlook*; FAO: **92**/Food and Agriculture Organization of the United States. In *Briefs: The State of World Fisheries and Aquaculture 2018. Meeting the Sustainable Development Goals*; **118** (bottom)/Food and Agriculture Organization of the United Nations, Reproduced with permission; **131**/© FAO 2019 http://www.fao.org/ save-food/resources/keyfindings/en/; **141** (top); **152** (bottom left); **234** (top)/© FAO Water use by five states in the Murray-Darling basin, 1920–2020 • Featherbrook: **178** (bottom)/Featherbrook • Geoimage Pty Ltd: **123** (left, right)/Data supplied and processed by Geoimage www.geoimage.com.au/Landsat 1987 • Geoscience Australia: **5, 55, 57, 84**/Commonwealth of Australia Geoscience Australia 2013. With the exception of the Commonwealth Coat of Arms and where otherwise noted, this product is provided under a Creative Commons Attribution 3.0 Australia Licence. • Getty Images Australia: **111**; **41** (top)/DEA/

S. VANNINI; **46**/© Ingetje Tadros; **69** (top)/UniversalImagesGroup; **244** (bottom)/Pallava Bagla/Corbis • Global Harvest Initiative: **137**/Re-drawn image from Global Harvest Initiative 2011 *GAP report: Measuring Global Agricultural Productivity*. Data from United Nations • Inter Press Service: **265** (bottom)/Photo by Jeffrey Lau, "Courtesy of IPS Inter Press Service" • IPCC: **126**/Figure SPM.7 from IPCC, 2014: *Climate Change 2014: Synthesis Report*. Contribution of Working Groups I, II and III to the Fifth Assessment Report of the Intergovernmental Panel on Climate Change, Figure SPM.11 upper panel. [Core Writing Team, Pachauri, R.K. and Meyer, L. eds.]. IPCC, Geneva, Switzerland.ps I, II and III to the Fourth Assessment Report of the Intergovernmental Panel on Climate Change [Core Writing Team, Pachauri, R.K. and Reisinger, A. eds.]. IPCC, Geneva, Switzerland. • ISAAA: **141** (bottom)/© ISAAA 2017. ISAAA. 2017. *Global Status of Commercialized Biotech/GM Crops in 2017: Biotech Crop Adoption Surges as Economic Benefits Accumulate in 22 Years*. ISAAA Brief No. 53. ISAAA: Ithaca, NY. pp.3 & 4 • John Wiley & Sons Australia: **9**, **91** (top), **193** (bottom), **200** (top), **203** (bottom), **217** (bottom), **259** (bottom); **192** (top, bottom)/Adapted from: Cosmetic Surgeon India and Rowena Ryan/News.com.Au; **196** (bottom)/Based on statistics from the UN World Tourism Organization • © JUMIA 2018: **254**, **258** (top, bottom) • MAPgraphics: **148**, **244** (top)/MAPgraphics • Mode Design: **63**/Emma-Jean Turner & David Bridgman • Mongabay: **82** (top, middle)/© Mongabay • NASA Earth Observatory: **101**, **82** (bottom); **119** (bottom left, bottom right)/NASA Earth Observatory images by Lauren Dauphin • National Geographic: **269**/Don Foley/National Geographic Stock • Ookla: **261** (top)/Source: Speedtest® by Ookla®. Analysis by Ookla of Speedtest Intelligence data February 2018. • Newspix: **64**/Justin Sanson; **160** (bottom)/Jono Searle • Out of Copyright: **59** • Panos Pictures: **44**, **236**; **73**/Aubrey Wade; **266** (top)/Andrew McConnell • Public Domain: **209** (top left), **245**; **86** (top)/Source: Euromonitor International; **96**/© European Union, 1995-2019. Cherlet, M., Hutchinson, C., Reynolds, J., Hill, J., Sommer, S.,von Maltitz, G. Eds., *World Atlas of Desertification*, Publication Office of the European Union, Luxembourg, 2018; **196** (top)/UN World Tourism Organization; **243** (top, bottom)/© World Bank; **250**/Regions are based on the ITU BDT Regions; **262**/Copyright © 2018. India Services , Ministry of Commerce and Industry, Government of India, India • Spatial Vision: **13** (top), **16**, **24**, **40** (top), **43** (top), **48**, **49**, **71**, **99** (bottom), **106**, **118** (top), **120**, **128** (bottom), **169** (c), **180**, **190** (bottom), **238**, **246**; **8** (top)/Data from *Reducing climate change impacts on agriculture: Global and regional effects of mitigation*, 2000–2080 by Tubiello F N, Fisher G in Technological Forecasting and Social Change 2007, 747: 1030-56. Map drawn by Spatial Vision; **51**/Map drawn by Spatial Vision, Source: © Commonwealth of Australia Geoscience Australia 2013; **80**/Data courtesy of the Institute on the Environment IonE, University of Minnesota. Map drawn by Spatial Vision; **100**/Data from Tony Burton. All rights reserved. Map drawn by Spatial Vision; **102**/Data from the USGS. Map drawn by Spatial Vision; **115**/© 2019 The Economist Intelligence Unit Limited , data from Global Food Security Index. Map drawn by Spatial Vision; **121**/Food and Agriculture Organisation, International Food Policy Research Institute; **195**/Data from World Tourism Organization. Map drawn by Spatial Vision.; **201**/Data © Commonwealth of Australia Geoscience Australia 2013 & © State of Queensland Department of Agriculture, Fisheries and Forestry 2013; **229**/Map drawn by Spatial Vision, data from WTO; **235**/Map drawn by Spatial Vision, data from Statista; Spatial Vision **240**/Map drawn by Spatial Vision. Data from © Commonwealth of Australia, DFAT, Australian Aid Budget Summary 2017–18; **124** (bottom)/Data from the Centre for Environmental Systems Research, University of Kassel. Map drawn by Spatial Vision; **128** (top)/Data from Reducing climate change impacts on agriculture: Global and regional effects of mitigation, 2000–2080 by Tubiello F N, Fisher G in Technological Forecasting and Social Change 2007, 747: 1030-56. Map drawn by Spatial Vision; **200** (bottom)/ABS, Austrade. Map drawn by Spatial Vision; **223** (left)/Data from Wikimedia Commons; **241** (bottom)/Map drawn by Spatial Vision. Data from UNDP Human Development Reports. • Redfin: **178** (top) • Rubbery Figures Pty Ltd: **108** (bottom)/Cartoon by Nicholson from The Australian • SecondBite: **149** (top) • Shutterstock: **1**/Triff; **2** (bottom)/Scott Prokop; **3** (top left)/Dmitri Ma; **3** (top centre)/Dmitry Kalinovsky; **3** (top right)/Toa55; **3** (bottom left)/avemario; **3** (bottom centre)/JB Manning; **3** (bottom right)/eakkachai/; **6**/Novikov Aleksey; **10** (bottom)/Sam Spicer; **10** (top)/Dudarev Mikhail; **12** (left)/Galyna Andrushko; **12** (right)/Dmitry Pichugin; **13** (bottom)/Evgeniya Moroz; **14** (top)/Kaesler Media; **14** (middle)/Eric Isselee; **14** (bottom)/Nicram Sabod; **15**/Vladimir Melnikov; **17**/Gbuglok; **20**/Snaprender; **21** (middle)/Marco Saracco; **21** (bottom)/totajla; **22** (top)/Richard

Whitcombe; **22** (middle)/kuehdi; **22** (bottom)/Janelle Lugge; **25** (top)/gillmar; **25** (bottom)/THPStock; **29** (right)/Catchlight Lens; **34** (top)/Joseph Sohm; **37**/Ashley Whitworth; **39**/Em7; **40** (bottom)/Hurst Photo; **41** (bottom)/CHEN WS; **47**/Sebastian Studio; **52** (bottom)/Orientaly; **53** (bottom)/Kaesler Media; **53** (top)/ Rosamund Parkinson; **54**/Phillip Minnis; **66**/Zzvet; **67**/John Bill; **69** (bottom)/smereka; **72**/tristan tan; **75**/erichon; **77**/pixinoo; **85** (left)/disfera; **85** (centre)/Fabio Berti; **85** (right)/Aaron Amat; **86** (bottom)/ David Hyde; **88**/Moreno Soppelsa; **91** (bottom left)/Andreas Altenburger; **91** (bottom right)/Anneka; **93**/Sukpaiboonwat; **95**/Dirk Ercken; **98**/Phillip Minnis; **105** (a)/Oleg Znamenskiy; **105** (b)/John Wollwerth; **105** (c)/Byelikova Oksana; **105** (e)/Michel Piccaya; **109** (top)/ Lockenes; **109** (bottom)/Dario Sabljak; **113**/R.M. Nunes; **114** (top)/I. Pilon; **114** (bottom)/Hector Conesa; **124** (top left)/Pataporn Kuanui; **134**/ChameleonsEye; **136**/CRSHELARE; **138**/Federico Rostagno; **139**/BRONWYN GUDGEON; **144** (flag)/Dana.S; **149** (bottom)/LittlePanda29; **158**/Hannamariah; **160** (top)/Anton Balazh; **162**/Zurijeta; **169** (a)/TK Kurikawa; **171**/Kummeleon; **172** (bottom)/Nils Versemann; **172** (top)/taewafeel; **173**/© Stephen Finn; **177** (bottom)/boreala; **182**/Richard Thornton; **185** (top)/tcly; **186**/Anton Balazh; **187**/goodluz; **191**/AVAVA; **194**/Lucky Business; **197**/iralu; **204** (left)/ChameleonsEye; **204** (right)/Villiers Steyn; **210**/Irina Silvestrova; **213** (top left)/Byelikova Oksana; **213** (top right)/BiksuTong; **214**/123Nelson; **215**/plavevski; **217** (top)/Anthony Ricci; **220**/Eugenie Photography; **222**/cozyta; **225** (top left)/Pressmaster; **225** (top centre)/bikeriderlondon; **225** (top right)/Roman Kosolapov; **225** (centre)/Pressmaster; **225** (bottom left)/Goodluz; **225** (bottom right)/Levent Konuk; **228** (bottom)/Claudine Van Massenhove; **230** (bottom)/Alf Manciagli; **237** (top)/Mohammad Saiful Islam; **247**/AlexLMX; **261** (bottom)/Noppasin Wongchum; **266** (bottom)/KPixMining; **271**/AlexRoz; **116**/© FAO, 2019. FAO, IFAD, UNICEF, WFP and WHO. 2018. *The State of Food Security and Nutrition in the World 2018. Building climate resilience for food security and nutrition.* Rome, FAO.Licence: CC BY-NC-SA 3.0 IGO. Map drawn by Spatial Vision • Sundrop Farms Pty Ltd: **142** • Tangaroa Blue Foundation: **35** • The Citizen: **175** • Turqle Trading South Africa: **237** (bottom) • UNWTO: **190** (top), **193** (top)/© UNWTO 2844/19/16 • Vanessa Harris: **163** • Viscopy: **165**/© Donkeyman Lee Tjupurrula/Licensed by Viscopy, 2016 • We Are Social Ltd: **251**/Sources: Population: United Nations: U.S. Census Bureau; Internet: InternetworldStats; ITU; Eurostat: Internetlivestats; CIA World Factbook: Mideastmedia.org; Facebook; Government Officaials; Regulatory Authories; Reputable Media; Social Media and Mobile Social Media: Facebook; Tencent; Vkontakte; Kakao; Naver; Ding; Techrasa; Similarweb; Kepios Analysis; Mobile: GSMA Intelligence; Google: Ericsson: Kepios Analysis. Note: Penetration figures are for total population all ages; **252**/Sources: Internetworldstats; ITU; Eurostat; Internetlivestats; CIA World Factbook; Mideastmedia.org; Facebook; Government Officials; Regulatory Authorities; Reputable Media. *Note:* Penetration Figures are for Total Population, regardless of age. • Wikimedia Commons: **29** (left), **31** (left), **31** (right), **60, 218** • WorldAtlas: **65**/WorldAtlas.com, http://www.worldatlas.com/articles/the-countries-producing-the-most-rice-in-the-world.html

Text

• Creative Commons: **45** (Table 1)/International Food Policy Research Institute; **152** (Table 1)/Copyright © 2016 International Center for Tropical Agriculture – CIAT; **226** (table)/Based on ABS trade data on DFAT STARS database, ABS catalogue 5368.0.55.003/4 and unpublished ABS data; **232** (table)/© Department of Foreign Affairs and Trade, *Trade and Investment at a glance* 2017 p. 27 • FAO: **45** (Table 2)/© FAO • FAPRI-ISU: **153** (table)/Dermot Hayes • Jason G. Goldman: **210** • Public Domain: **41**/© The Daily Records. Top 12 largest maize producing countries in the world. By Abayomi Jegede, January 1, 2019; **144** (Figure 1 text)/© World Food Programme 2018. • UN World Tourism Organization: **212** • We Are Social Ltd: **252–3** (table)/Sources: Internetworldstats; ITU; Eurostat: InternetLivestats; CIA World Factbook; Mideastmedia.org; Facebook; Government Officials; Regulatory Authorities; Reputable Media. *Note:* Penetration figures are for total population, regardless of age.

Every effort has been made to trace the ownership of copyright material. Information that will enable the publisher to rectify any error or omission in subsequent reprints will be welcome. In such cases, please contact the Permissions Section of John Wiley & Sons Australia, Ltd.

1 Geographical skills and concepts

1.1 Overview

1.1.1 Introduction

As a student of Geography, you are building knowledge and skills that will be needed by you and your community now and into the future. The concepts and skills that you use in Geography can also be applied to everyday situations, such as finding your way from one place to another. Studying Geography may even help you in a future career here in Australia or somewhere overseas.

Throughout your study of Geography, you will cover topics that will give you a better understanding of the social and physical aspects of the world around you, at both the local and global scale. You will investigate issues that need to be addressed now and in the future.

LEARNING SEQUENCE

1.1 Overview
1.2 Work and careers in Geography
1.3 Concepts and skills used in Geography
1.4 **Review**

To access interactivities and resources, select your learnON format at www.jacplus.com.au.

1.2 Work and careers in Geography

1.2.1 Skills for work

In Geography, students develop an understanding of the world. These skills are transferable to the workplace and can be used as a basis for evaluating strategies for the sustainable use and management of the world's resources. An understanding of Geography and its application for managing sustainable futures is pivotal knowledge that will be desirable to many future employers.

Geographical skills and knowledge are a foundation for many occupations. The study of Geography includes developing important geospatial and spatial technology skills, which underpin the knowledge base of a range of courses and careers.

- *Geospatial skills:* the ability to collect and collate information gathered from fieldwork and observations. Geospatial skills are used in careers such as surveying, meteorology, agricultural science and urban planning.
- *Spatial technologies*: technologies that demonstrate the connections between location, people and activities in digital formats. Jobs in the spatial industry are varied and include working in business and government. Spatial technologies apply many techniques, such as photogrammetry, remote sensing and global positioning systems (GPS). Spatial technologies manage information about the environment, transportation and other utility systems.

FIGURE 1 Using GPS to survey and record road traffic movements for a local council

FIGURE 2 GIS (geographical information systems) being used to manage spaces and plan escape routes during a fire

1.2.2 Where can Geography lead?

Careers that draw on Geography as a foundation skill are many and varied. As you consider your pathway options for senior studies you may like to research some of the careers that are provided in **FIGURE 3**.

FIGURE 3 Geography pathways

Meteorologist	**Surveyor**	**Landscape architect**
Meteorologists use geographical skills to forecast the weather, study the atmosphere and understand climate change.	Surveyors use geographical skills to measure, analyse and report on land-related information for planning and development.	Landscape architects use geographical skills to plan and design land areas for large-scale projects such as housing estates, schools, hospitals and gardens.

Agricultural technician	**Park ranger**	**Environmental manager**
Agricultural technicians use geographical skills to advise farmers on aspects of agriculture such as crop yield, farming methods, production and marketing.	Park rangers use geographical skills to support and maintain ecosystems in national parks, scenic areas, historic sites, nature reserves and other recreational areas.	Environmental managers use geographical skills for project management and the development of environmental reports.

on Resources

🔗 **Weblink** Job Outlook

1.2 INQUIRY ACTIVITY

Job Outlook is a federal government website that provides information on employment in a range of occupations. It also includes information on the training, skills and tools needed for various careers. Use the **Job Outlook** weblink in the Resources tab to complete the following.

a. Select one of the occupational profiles presented in **FIGURE 3** that interests you and use the **Job Outlook** weblink to explore and learn more about this career.

b. Develop a career profile for your occupation of choice. In your profile include:
- the geographical skills needed for this job
- the geographical tools that may be used in this occupation
- the study and training requirements that lead to this occupation
- the job prospects for your chosen occupation over the next five years.

Examining, analysing, interpreting

1.3 Concepts and skills used in Geography

1.3.1 Skills used in studying Geography

As you work through each of the topics in this title, you'll complete a range of exercises to check and apply your understanding of concepts covered. In each of these exercises, you'll use a variety of skills, which are identified using the Geographical skills (GS) key provided at the start of each exercise set. These are:

- **GS1** Remembering and understanding
- **GS2** Describing and explaining
- **GS3** Comparing and contrasting
- **GS4** Classifying, organising, constructing
- **GS5** Examining, analysing, interpreting
- **GS6** Evaluating, predicting, proposing

In addition to these broad skills, there is a range of essential practical skills that you will learn, practise and master as you study Geography. The SkillBuilder subtopics found throughout this title will tell you about the skill, show you the skill and let you apply the skill to the topics covered.

The SkillBuilders you'll use in Year 9 are listed below.

- Describing spatial relationships in thematic maps
- Constructing and describing a transect on a topographic map
- Constructing ternary graphs
- Describing patterns and correlations on a topographic map
- GIS — deconstructing a map
- Interpreting a geographical cartoon
- Constructing and describing complex choropleth maps
- Interpreting satellite images to show change over time
- Constructing a box scattergram
- Constructing and describing proportional circles on maps
- Interpreting topological maps
- Constructing and describing isoline maps
- Constructing and describing a doughnut chart
- Describing divergence graphs
- Constructing multiple line and cumulative line graphs
- Constructing and describing a flow map
- Constructing a table of data for a GIS
- Using advanced survey techniques — interviews

1.3.2 SPICESS

Geographical concepts help you make sense of your world. By using these concepts you can investigate and understand the world you live in, and you can use them to try to imagine a different world. The concepts help you to think geographically. There are seven major concepts: *space*, *place*, *interconnection*, *change*, *environment*, *sustainability* and *scale*. We will explore each of these concepts in detail in the following sections and through the activities and exercises in this subtopic.

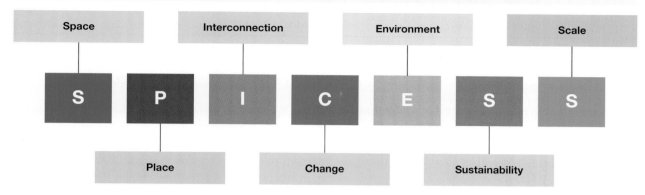

FIGURE 1 A way to remember the seven geographical concepts is to think of the term SPICESS.

Space

Interconnection

Environment

Scale

S P I C E S S

Place

Change

Sustainability

1.3.3 What is space?

Everything has a location on the space that is the surface of the Earth. Studying the effects of location, the distribution of things across this space, and how the space is organised and managed by people, helps us to understand why the world is like it is.

A place can be described by its absolute location: for example, latitude and longitude, a grid reference, street directory reference or an address. Or, a place can be described using its relative location — where it is in relation to another place in terms of distance and direction.

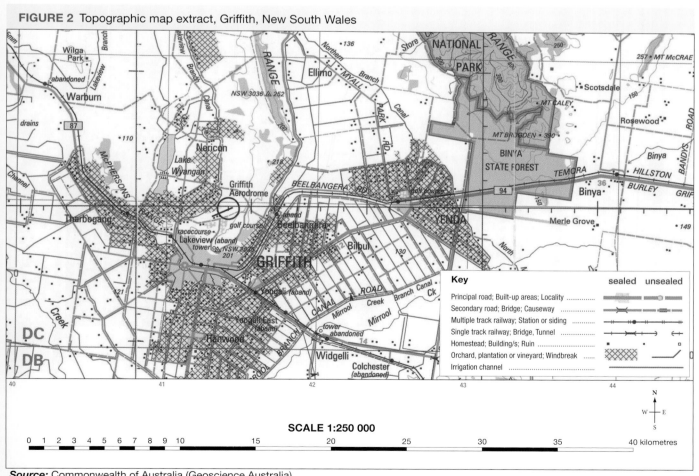

FIGURE 2 Topographic map extract, Griffith, New South Wales

Source: Commonwealth of Australia (Geoscience Australia)

1.3.4 What is place?

The world is made up of places, so to understand our world we need to understand its places by studying their variety, how they influence our lives and how we create and change them.

Everywhere is a place. Each of the world's biomes — for example, a desert environment — can be considered a place, and within each biome there are different places, such as the Sahara Desert. There can be natural places — an oasis is a good example — or man-made places such as Las Vegas. Places can have different functions and activities — for example, Canberra is an administration centre, while the MCG is a place for major sporting events and the Great Barrier Reef is a place of great natural beauty with a coral reef biome. People are interconnected to places and other people in a wide variety of ways — for example, when we move between places or connect electronically via computers. We are connected to the places that we live in or know well, such as our neighbourhood or favourite holiday destination.

FIGURE 3 Located in a desert biome, this array of greenhouses in Almeria, Spain, allows for the control of soil, moisture, nutrients and weather conditions, enabling the large-scale farming of fruit and vegetables.

1.3.5 What is interconnection?

People and things are connected to other people and things in their own and other places, and understanding these connections helps us to understand how and why places are changing.

Individual geographical features can be interconnected — for example, the climate within a place or biome, such as a tropical rainforest, can influence natural vegetation, while removal of this vegetation can affect climate. People can be interconnected to other people and other places via employment, communications, sporting events or cultural ties. The manufacturing of a product may create interconnections between suppliers, manufacturers, retailers and consumers. Trade in goods and services creates interconnections across the globe.

FIGURE 4 World trade flows — exports of agricultural products by region, US$ billion

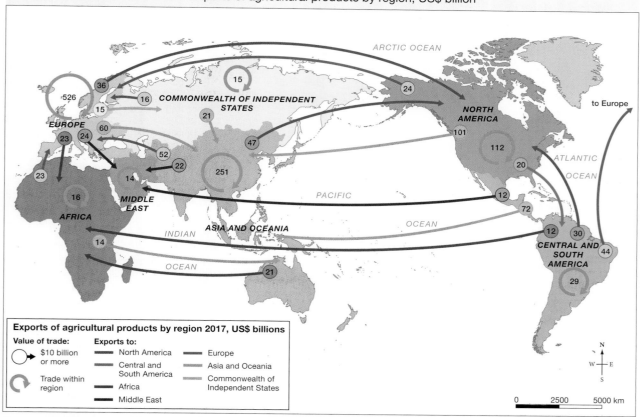

Source: Data from World Trade Organization.

┌─ Explore more with my**World**Atlas ─────────────────────────────

Deepen your understanding of this topic with related case studies and questions.
• Developing Australian Curriculum concepts > **Interconnection**

1.3.6 What is change?

The concept of change is about using time to better understand a place, an environment, a spatial pattern or a geographical problem.

From a geographical time perspective, change can be very slow — think of processes such as the formation of mountains or soil. On the other hand, a volcanic eruption or landslide can change landforms rapidly. It may take some years for the boundary of a city to expand outwards, but in the space of a few weeks whole suburbs can be demolished to make way for a freeway. Change can also have physical, economic and social implications. Consider the effect of the internet over the past few years.

FIGURE 5 Predictions of the effects of climate change on cereal crops

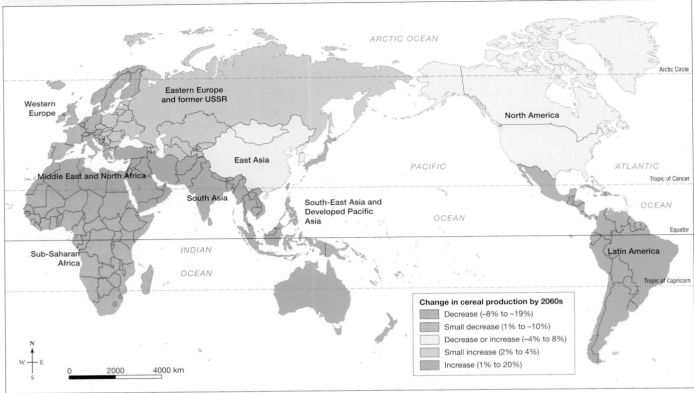

Change in cereal production by 2060s

- Decrease (–8% to –19%)
- Small decrease (1% to –10%)
- Decrease or increase (–4% to 8%)
- Small increase (2% to 4%)
- Increase (1% to 20%)

Source: Spatial Vision

Explore more with my**World**Atlas

Deepen your understanding of this topic with related case studies and questions.
- Developing Australian Curriculum concepts > **Change**

1.3.7 What is environment?

People live in and depend on the environment, so it has an important influence on our lives.

The biological and physical world that makes up the environment is important to us as a source of food and raw materials, a means of absorbing and recycling wastes, and a source of enjoyment and inspiration.

People perceive, adapt and use environments in many ways. For example, different people could look at a well-vegetated hillside and one might see it as a source of timber for construction, another might see a slope that could be cleared and terraced to produce food, while another might view it as a scenic environment for ecotourism.

FIGURE 6 The East Kolkata wetlands act as a sewage filtration system and recycle nutrients through the soil to allow a wide range of food crops to be grown. The ponds provide one-third of the city's fish supply and are a protected Ramsar site for migratory birds.

1.3.8 What is sustainability?

Sustainability is about maintaining the capacity of the environment to support our lives and those of other living creatures.

Sustainability involves maintaining and managing our resources and environments for future generations. It is important to understand the causes of unsustainable situations to be able to make informed decisions on the best way to manage our natural world.

FIGURE 7 The unsustainable nature of fishing

1.3.9 What is scale?

When we examine geographical questions at different spatial levels we are using the concept of scale to find more complete answers.

A little like a zoom lens on a camera, scale enables us to examine issues from different perspectives, from personal to local, regional, national or global. Using scale helps in the analysis and explanation of phenomena. For example, climate is the most important factor in determining vegetation type on a global scale, whereas, at a local scale, soil and drainage might be more important. Different activities can also have an impact at a range of scales; for example, the construction of an international airport in Cairns saw the development of tourism evolve from a local to an international scale, with direct flights between Australia and South-East Asia.

FIGURE 8 Country of origin for tourists visiting Australia, and destinations for Australian tourists

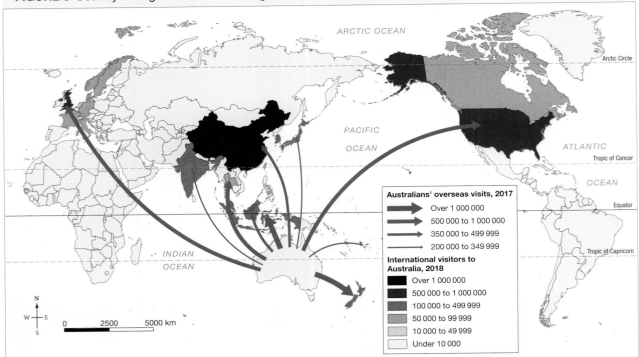

Source: ABS, Austrade.

Explore more with my**World**Atlas

Deepen your understanding of this topic with related case studies and questions.
• Developing Australian Curriculum concepts > **Scale**

1.3 ACTIVITIES

1. Create a diagram to show the ***interconnections*** that could occur for the growing, manufacturing, sales and consumption of a can of pineapple slices. **Classifying, organising, constructing**
2. (a) Brainstorm with your class examples of ***environmental*** issues at local, national and global ***scales***.
   (b) How would solutions differ at each ***scale***? **Evaluating, predicting, proposing**

1.4 Review

on**line** only

1.4.1 Key knowledge summary
Use this dot point summary to review the content covered in this topic.

1.4 Exercise 1: Review
Select your learnON format to complete review questions for this topic.

 Resources

 eWorkbook Crossword (doc-31713)

 Interactivity Geographical skills and concepts crossword (int-7643)

UNIT 1
BIOMES AND FOOD SECURITY

Food is essential to human life. To ensure we have reliable food sources, we alter our world biomes — clearing vegetation, diverting and storing water, adding chemicals and even changing landforms. We will need to carefully manage our limited land and water resources and use more sustainable farming practices to ensure we have future food security.

GEOGRAPHICAL INQUIRY: BIOMES AND FOOD SECURITY

Your task

Your team has been selected to create a website that not only grabs people's attention but also informs them of the importance of one particular biome as a producer of food, and the current threats to food production. Looking into the future, you will also suggest more sustainable ways of managing this biome.

Select your learnON format to access:

- an overview of the project task
- details of the inquiry process
- resources to guide your inquiry
- an assessment rubric.

Resources

ProjectsPLUS   Geographical inquiry: Biomes and food security (pro-0148)

2 All the world is a biome

2.1 Overview

What on Earth are biomes? Are they just another part of the landscape or do we need them to survive?

2.1.1 Introduction

Where do the foods we eat and the natural products we use daily come from? The answer is biomes. Biomes are communities of plants and animals that extend over large areas. Some are dense forests; some are deserts; some are grasslands, like much of Australia; and so the variations continue. Within each biome, plants and animals have similar adaptations that allow them to survive.

Biomes can be terrestrial (land based) or aquatic (water based). Understanding the diversity within them is essential to our survival and wellbeing.

Within each biome, there are many variations in the landscape and climate, and in the plants and animals that have adapted to survive there.

on Resources

  eWorkbook   Customisable worksheets for this topic

  Video eLesson Bountiful biomes (eles-1717)

To access a pre-test and starter questions and receive immediate, **corrective feedback** and **sample responses** to every question, select your learnON format at www.jacplus.com.au.

2.2 Defining biomes

2.2.1 What and where are the major biomes?

Biomes are sometimes referred to as ecosystems. They are places that share a similar climate and life forms. There are five distinct biomes across the Earth: forest, desert, grassland, tundra and aquatic biomes. Within each, there are variations in the visible landscape, and in the plants and animals that have adapted to survive in a particular climate.

FIGURE 1 Major biomes of the world

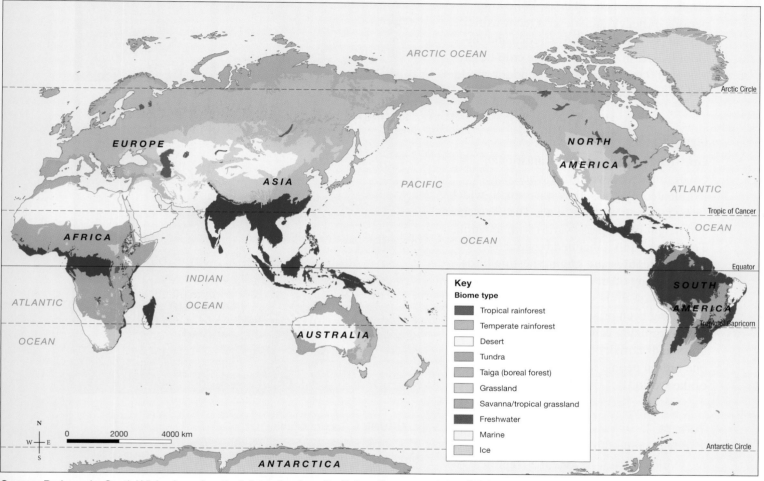

Key
Biome type
- Tropical rainforest
- Temperate rainforest
- Desert
- Tundra
- Taiga (boreal forest)
- Grassland
- Savanna/tropical grassland
- Freshwater
- Marine
- Ice

Source: Redrawn by Spatial Vision based on the information from the Nature Conservancy and GIS Data

Forests

Forests are the most diverse ecosystems on the Earth. Ranging from hot, wet, tropical rainforests to temperate forests, they have an abundance of both plant and animal life. Over 50 per cent of all known plant and animal species are found in tropical rainforests. Forests are the source of over 7000 modern medicines, and many fruits and nuts originated in this biome. Forests help regulate global climate, because they absorb and use energy from the sun rather than reflect it back into the atmosphere. Forest plants recycle water back into the atmosphere, produce the oxygen we breathe, and store the carbon we produce. Forests are under threat from **deforestation**.

FIGURE 2 Forest biome

Deserts

Deserts are places of low rainfall and comprise the arid and semi-arid regions of the world. Generally they are places of temperature extremes — hot by day and cold by night. Most animals that inhabit deserts are nocturnal (active at night), and desert vegetation is sparse. Desert rain often evaporates before it hits the ground, or else it falls in short, heavy bursts. Following periods of heavy rain, deserts teem with life. Not all deserts are hot. Antarctica and the Gobi Desert in central Asia are cold deserts. About 300 million people around the world live in desert regions.

Grasslands

Grasslands can be seen as transitional environments between forest and desert. Dominated by grass, they have small, widely spaced trees or no trees at all. The coarseness and height of the grass varies with location. They are mainly inhabited by grazing animals, reptiles and ground-nesting birds, though many other animals can be found in areas with more tree cover. Grasslands have long been prized for livestock grazing, but overgrazing is unsustainable and places grasslands at risk of becoming deserts. Over one billion people inhabit the grassland areas of the world.

Tundra

Tundra is found in the coldest regions of the world, and lies beyond the **treeline**. The landscape is characterised by grasses, dwarf shrubs, mosses and lichens. The growing season is short. Tundra falls into three distinct categories — Arctic, Antarctic and alpine — but they share the common characteristic of low temperatures. In Arctic regions there is a layer beneath the surface known as permafrost — permanently frozen ground. The tundra biome is the most vulnerable to global warming, because its plants and animals have little tolerance for environmental changes that reduce snow cover.

FIGURE 3 Desert biome

FIGURE 4 Grassland biome

FIGURE 5 Tundra biome

Aquatic biomes

Water covers about three-quarters of the Earth. Aquatic biomes can be classified as freshwater or marine. Freshwater biomes contain very little salt and are found on land; these include lakes, rivers and wetlands. Marine biomes are the saltwater regions of the Earth and include oceans, coral reefs and estuaries. Marine environments are teeming with plant and animal life, and are a major food source. Compounds from marine life have also been used in products such as cosmetics and toothpaste. Elements taken from the roots of mangroves have been used in the development of cancer medications.

FIGURE 6 Aquatic biome

 **Resources**

 **Interactivity** Beautiful biomes (int-3317)

2.2 EXERCISES

Geographical skills key: GS1 Remembering and understanding **GS2** Describing and explaining **GS3** Comparing and contrasting **GS4** Classifying, organising, constructing **GS5** Examining, analysing, interpreting **GS6** Evaluating, predicting, proposing

2.2 Exercise 1: Check your understanding

1. **GS1** Name the five major biomes of the Earth and classify them as either aquatic or terrestrial.
2. **GS1** Identify the broad characteristics shared by biomes.
3. **GS2** Look carefully at **FIGURE 1**. Using geographic terminology and concepts (including reference to latitude), describe the location and characteristics of the major biomes.
4. **GS2** Are all biomes equally important? Explain your answer.
5. **GS1** Which biome has the greatest biodiversity?

2.2 Exercise 2: Apply your understanding

1. **GS2** Explain the important functions performed by the forest biome.
2. **GS6** Select one of the categories of biomes described in this subtopic. Suggest how this biome might be **changed** and used by humans and what impact this **change** might have on the **environment**.
3. **GS2** With the aid of a diagram, explain how forests help regulate global climate.
4. **GS6** Predict what might happen if the permafrost beneath the Arctic surface thawed.
5. **GS6** 'Deserts are a dry, lifeless plain.' To what extent do you agree with this statement? Explain.

Try these questions in learnON for instant, corrective feedback. Go to www.jacplus.com.au.

2.3 SkillBuilder: Describing spatial relationships in thematic maps

What are spatial relationships in thematic maps?

A spatial relationship is the interconnection between two or more pieces of information in a thematic map, and the degree to which they influence each other's distribution in space. Describing these relationships helps us understand how one thing affects another.

Select your learnON format to access:

- an explanation of the skill (Tell me)
- a step-by-step process to develop the skill, with an example (Show me)
- an activity to allow you to practise the skill (Let me do it)
- questions to consolidate your understanding of the skill.

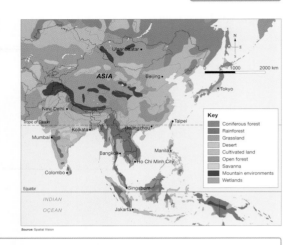

on Resources

Video eLesson Describing spatial relationships in thematic maps (eles-1726)

Interactivity Describing spatial relationships in thematic maps (int-3344)

2.4 The characteristics of biomes

2.4.1 Climate's influence on biomes

Biomes are controlled by climate. In turn, climate is influenced by factors such as the distance from the equator, altitude and distance from the sea, the direction of prevailing winds, and the location of mountain ranges. These play a key role in determining a region's climate and soil, which ultimately influence which plants and animals will inhabit it.

Temperature and rainfall patterns across the Earth determine which plant and animal species can survive in a particular biome. For instance, a polar bear could not survive in the hot climate of a desert or a tropical rainforest. Camels would not survive in the polar regions of the Earth, and fish cannot survive without water. The plants and animals of a region have adapted over time to the variations in the region's climate conditions.

2.4.2 What influences climate?

The geographical features of the Earth's surface, such as mountain ranges and **latitude**, are key influences on climate. Similarities have been found in the adaptations of plant and animal species in mountain regions and those found near the poles.

Landform

The major geographical influence on climate is the location of mountain ranges (see **FIGURE 1**). Mountain ranges affect the amount of **precipitation** that reaches inland areas, because they pose a barrier to moisture-laden prevailing winds. **Rain shadows** form on the **leeward** side of mountains (opposite to the **windward** side). Deserts often form in rain shadows.

FIGURE 1 The influence of mountains on climate. This illustration shows the pattern typical on the east coast of Australia, where there are warm ocean currents.

Rising moist air produces rain.

Dry air continues over mountains.

Winds become dry by the time they reach inland areas.

Trade winds are forced to rise.

Inland

Sea

Coast

Desert

Mountains

Thousands of kilometres

FIGURE 2 Mt Kilimanjaro is only three degrees south of the equator but it is 5895 metres high; its altitude is the reason it has snow on its summit.

Altitude also plays a significant role in determining the climate. Temperatures fall by 0.65 °C for every 100-metre increase in elevation. This can be illustrated by Mt Kilimanjaro (**FIGURE 2**), which is located on the border of Tanzania and Kenya, in Africa, at approximately 3° latitude from the equator. Towering 5895 metres above sea level, Mt Kilimanjaro is the highest mountain in Africa. Depending on the time of the day, the temperature at the base of the mountain ranges from 21 °C to 27 °C. At the summit, temperatures can plummet to −26 °C. As you move from base to summit, variations occur in the landscape as it transitions from rainforest to alpine desert to desert tundra.

Latitude

The sun's rays are more direct at the equator. With more energy focused on that region, it heats up more quickly. At the poles, the sun's rays are spread over a larger area and therefore cannot heat up as effectively. As a result, areas at the poles are much cooler than areas at the equator (see **FIGURE 3**).

The tilt of the Earth on its axis also has a role to play. When a hemisphere tilts towards the sun, the sun's rays hit it more directly. This means that a larger space is in more intense sunlight for longer. The days are longer and warmer, and the hemisphere experiences summer. The reverse is true when a hemisphere tilts away from the sun in winter.

Ocean currents and air movement

There are other factors that influence climate and play a role in the development of biomes. Two of these are ocean currents and air movement. In addition, differences also occur when you move from the coast to inland areas.

When cold ocean currents flow close to a warm land mass, a desert is more likely to form. This is because cold ocean currents cool the air above, causing less evaporation and making the air drier. As this air moves over the warm land, it heats up, making it less likely to release any moisture it holds; thus, deserts form. For example, cold ocean currents flow off the coast of Western Australia, while the east-coast Pacific Ocean currents are warmer. As a result, Perth on average receives less rainfall than Sydney.

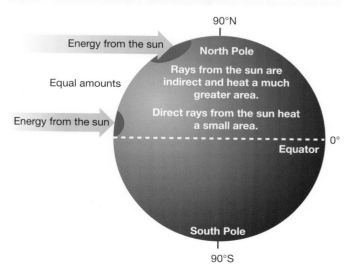

FIGURE 3 The influence of latitude on climate. The rotation of the Earth around the sun and the tilt of the Earth on its axis also influence the seasons.

90°N

Energy from the sun

North Pole

Rays from the sun are indirect and heat a much greater area.

Equal amounts

Direct rays from the sun heat a small area.

Energy from the sun

Equator   0°

South Pole

90°S

2.4.3 The role of soil in biomes

Soil is important in determining which plants and animals inhabit a particular biome. Soils not only vary around the world but also within regions. The characteristics of soil are determined by:

- temperature
- rainfall
- the rocks and minerals that make up the bedrock, which is the basis of soil development.

The amount of vegetation present also plays an important role in determining the quality of the soil. **FIGURE 4** shows a typical soil profile. The different soil layers are referred to as horizons.

Why do soils differ?

Biomes located in the high latitudes (those farthest from the equator) have lower temperatures and less exposure to sunlight than biomes located in the low latitudes (those close to the equator). There are also variations in the amount of precipitation that biomes receive. This is determined partly by their location in relation to the equator: lower latitude regions generally receive more precipitation than those in higher latitudes (see **FIGURE 5**).

Temperature and precipitation patterns are important factors in determining the rate of soil development. In addition, soil moisture, its nutrient content and the length of the growing season also play key roles in soil development and, ultimately, the biodiversity of a biome.

Soil is more abundant in biomes that have both high temperatures and high moisture than in cold, dry regions. This is because erosion of bedrock is more rapid when moisture content is high, and organic material decomposes at a faster rate in high temperatures. The decomposition of organic matter provides the nutrients needed for plant growth, which in turn die and decompose in a continuous cycle. This is further demonstrated in **FIGURE 6**.

FIGURE 4 A typical soil profile has a number of distinct layers.

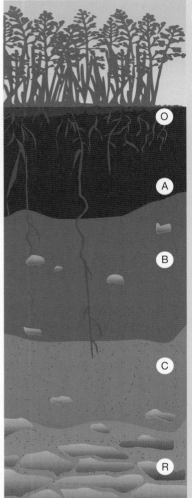

Horizon O (organic matter): A thin layer of decomposing matter, humus, and material that has not started to decompose, such as leaf litter

Horizon A (topsoil): The upper layer of soil, nearest the surface. It is rich in nutrients to support plant growth and usually dark in colour. Most plant roots and soil organisms are found in this horizon, which will also contain some minerals. In areas of high rainfall, such as tropical rainforests, minerals will be leached out of this layer. A constant supply of decomposing organic matter is needed to maintain soil fertility.

Horizon B (subsoil): Plant litter is not present in horizon B; as a result, little humus is present. Nutrients leached from horizon A accumulate in this layer, which will be lighter in colour and contain more minerals than the horizon above.

Horizon C (parent material): Weathered rock that has not broken down far enough to be soil. Nutrients leached from horizon A are also found in this layer. It will have a high mineral content; the type is determined by the underlying bedrock.

Horizon R (bedrock): Underlying layer of partly weathered rock

FIGURE 5 Latitude is a key factor in climate, which in turn is linked to soil characteristics. Soils in high-latitude biomes differ from those in low latitudes.

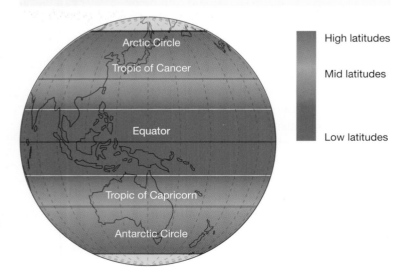

High latitudes

Mid latitudes

Low latitudes

FIGURE 6 Different biomes have different soil and vegetation characteristics.

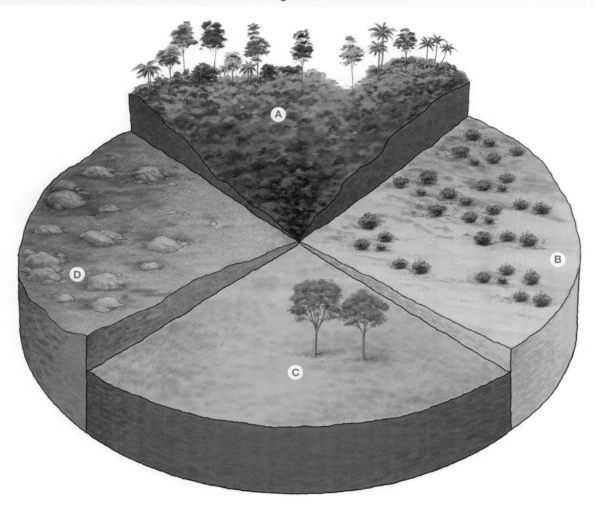

(A) **Tropical rainforest**

- High temperatures cause weathering, or breakdown, of rocks and organic matter.
- High rainfall **leaches** nutrients from the soil.
- Soil is often reddish because of high iron levels.
- Organic matter is often a shallow layer on the surface. Nutrients are constantly recycled, allowing the rainforest to flourish.
- Soil fertility is rapidly lost if trees are removed, as the supply of organic material is no longer present.

(B) **Desert**

- Limited vegetation means a limited supply of organic material for soil development.
- High temperatures rapidly break down any organic material.
- Soils are pale in colour rather than dark.
- Lack of rainfall limits plant growth.
- Lack of vegetation makes surface soil unstable and easily blown away.
- Soil does not have time to develop and mature.

(C) **Temperate**

- Generally brown in colour, soils have distinctive horizons and are generally about a metre deep.
- Soils are ideal for agriculture; they are not subjected to the extremes of climate found in high and low latitudes.
- Moderate climate; temperature and rainfall are sufficient for plant growth.
- Dominated by temperate grasslands and deciduous forests.

(D) **Tundra**

- Soil is shallow and poorly developed.
- Some layers are frozen for long periods.
- Subsoil may be permanently frozen.
- Soil is covered by ice and snow for most of the year.
- Growing season may be limited to a few weeks.
- Soil may contain large amounts of organic material but extreme cold means it breaks down very slowly.
- Trees are absent; mosses and stunted grasses dominate.

What else is in the soil?

Soil not only supports the plants and animals that we see on the surface of the land; the soil itself is also home to a variety of life forms such as bacteria, fungi, earthworms and algae.

While most soil organisms are too small to be seen, there are others that are visible. For instance, more than 400 000 earthworms can be found on a hectare of land. Regardless of size, all soil organisms play a vital role in maintaining soil quality and fertility. For example, earthworms:

- compost waste and fertilise the soil
- improve drainage and aeration
- bring subsoil to the surface and mix it with topsoil
- secrete nitrogen and chemicals that help bind the soil.

FIGURE 7 There are more microbes in a teaspoon of soil than there are people on Earth.

 Resources

 Interactivity Why are biomes different? (int-3319)

2.4 INQUIRY ACTIVITY

Use your atlas to locate Rwanda in central Africa.
a. What type of biome would you expect to find in Rwanda? Give reasons for your answer.
b. What do you think the soil would be like in Rwanda? Use the internet to test your theory.

Evaluating, predicting, proposing

2.4 EXERCISES

Geographical skills key: GS1 Remembering and understanding **GS2** Describing and explaining **GS3** Comparing and contrasting **GS4** Classifying, organising, constructing **GS5** Examining, analysing, interpreting **GS6** Evaluating, predicting, proposing

2.4 Exercise 1: Check your understanding

1. **GS1** What are the major influences on the development of biomes?
2. **GS2** Explain the difference between the windward and leeward side of a mountain range.
3. **GS2** Explain how temperature and rainfall can influence the development of soil across different biomes.
4. **GS2** Outline the different characteristics of soils in the tropical rainforest, tundra, desert and temperate biomes.
5. **GS2** Describe the role played by soil organisms in maintaining soil quality and fertility.

2.4 Exercise 2: Apply your understanding

1. **GS6** Would you expect to find soil variations within biomes? Give reasons for your answer.
2. **GS6** Predict the **changes** that might occur if earthworms or micro-organisms within the soil were no longer present.
3. **GS6** What type of climate and biomes would you expect to find at the equator? Why?
4. **GS5** Describe the climate and landscape on Mount Kilimanjaro. Why is there so much variation?
5. **GS5** What are some of the factors that create variations in biomes?

Try these questions in learnON for instant, corrective feedback. Go to www.jacplus.com.au.

2.5 Australia's major biomes

2.5.1 What factors shape Australian biomes?

Australia is a land of contrasts. Its mountain ranges and river systems are small by world standards. In the north there are tropical rainforests and savanna grasslands. In the centre there is a wide expanse of desert that is second in area only to the Sahara Desert in Africa. In the south, temperate forests and grasslands dominate. Australia also has vast wetlands and coastal ecosystems.

Before European colonisation, the Australian landscape was shaped largely by natural processes and Aboriginal burning practices. With European occupation came large-scale land clearing, irrigation of the land through water diversion from rivers, and the draining of wetlands. New plant and animal species were introduced to the landscape. However, despite the large-scale changes made since European occupation, Australia's major biomes are still clearly evident.

FIGURE 1 Climate classification of Australia

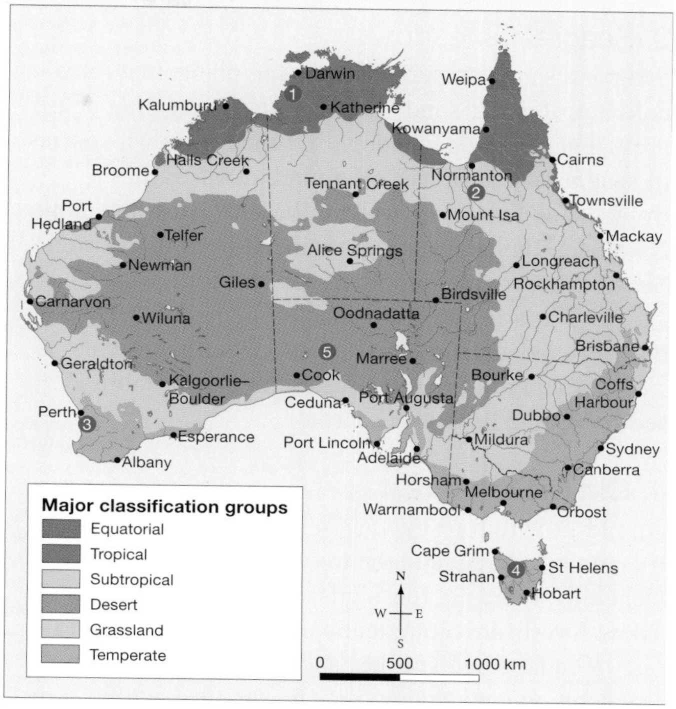

Major classification groups
- Equatorial
- Tropical
- Subtropical
- Desert
- Grassland
- Temperate

Source: Data copyright Commonwealth of Australia, 2013 Bureau of Meteorology. Map drawn by Spatial Vision.

1. Wetlands and rivers

In northern Australia, wetlands have been inhabited by Indigenous Australian peoples since the beginning of the Dreaming (more than 50 000 years). These areas provided them with food and water, and they used wetland plants such as river reeds and lily leaves to make fishing traps. Today, wetlands are still important habitats for native and migratory birds. In many parts of Australia, they are under threat, because water is diverted from rivers to produce food crops and cotton.

FIGURE 2 This billabong in Kakadu National Park is part of an extensive wetland system that develops during the wet season.

2. Savanna (grasslands)

Grasslands (or savanna) are generally flat, with either few trees and shrubs, or very open woodland. For many native species, grasslands provide vital habitat and protection from predators. Many grasslands depend on a regular cycle of burning to germinate their seeds and to revive the land. Periodic burning also prevents trees from gaining dominance in the landscape. Before European occupation, Indigenous Australian peoples hunted the animals in the grasslands. However, since then, grasslands have been used extensively for grazing.

FIGURE 3 Savanna biomes are typically dominated by grasses and scattered trees.

These areas often mark the transition between desert and forest, and are a very fragile biome. Without careful management they can quickly change to desert. Less than one per cent of Australia's original native grasslands survive today.

3. Seagrass meadows

Seagrasses are submerged flowering plants that form colonies off long, sandy ocean beaches, creating dense areas that resemble meadows. Of the 60 known species of seagrass, at least half are found in Australia's tropical and temperate waters. Western Australia alone is home to the largest seagrass meadow in the world. Seagrasses provide important habitats for a wide variety of marine creatures, including rock lobsters, dugongs and sea turtles. They also absorb nutrients from coastal run-off, slow water flow, help stabilise sediment, and keep water clear.

FIGURE 4 Seagrass meadows provide food, shelter and breeding grounds for marine life.

4. Old-growth forest

An old-growth forest is one in its oldest growth stage. It is multi-layered, and the trees are of mixed ages. Generally, there are few signs of human disturbance. These forests are biologically diverse, often home to rare or endangered species, and show signs of natural regeneration and decomposition. The trees within some old-growth forests have been felled for their timber and to create paper products. **Logging** can reduce **biodiversity**, affecting not only the forest itself but also the indigenous plant and animal species that rely on the old-growth habitat. It is estimated that **clear-felling** of Tasmania's old-growth forests would release as much as 650 tonnes of carbon per hectare into the atmosphere. In Victoria, near Melbourne, many old-growth forests lie within protected water supply catchments and help maintain the integrity of the city's water supply.

FIGURE 5 Different layers of vegetation can be seen in old-growth forests.

5. Desert

Australian deserts are places of temperature extremes. During the day, temperatures sometimes exceed 50 °C, but at night this can drop to freezing. Australia's desert regions are often referred to as the outback but they are not all endless plains of sand. Some, such as the Simpson Desert and the Great Sandy Desert, are dominated by sand. The Nullarbor Plain and Barkly Tableland are mainly smooth and flat, while the Gibson Desert and Sturt Stony Desert contain low, rocky hills. In some areas, the landscape is dominated by spinifex and acacia shrubs.

FIGURE 6 Vegetation in desert biomes has specific adaptations that enable it to survive in the harsh climate.

┌─ Explore more with my**World**Atlas ───

Deepen your understanding of this topic with related case studies and questions.
• Investigating Australian Curriculum topics > Year 9: Biomes and food security >**Australia's alpine biomes**

2.5 INQUIRY ACTIVITIES

1. Select one of the climate zones shown in **FIGURE 1** and investigate the biomes found within it. Prepare a report on the importance of one of these biomes and discuss how it has *changed* over time. What do you think should be done to protect it?   **Examining, analysing, interpreting**
2. (a) Investigate one of the Australian biomes and examine how plants and animals have adapted to survive in it. Create a class collage depicting the way plants and animals have adapted to the Australian *environment*.
   (b) Explore what this biome is like in another *place* on Earth. With the aid of a Venn diagram, compare the two *places*.   **Examining, analysing, interpreting**

2.5 EXERCISES

Geographical skills key: GS1 Remembering and understanding **GS2** Describing and explaining **GS3** Comparing and contrasting **GS4** Classifying, organising, constructing **GS5** Examining, analysing, interpreting **GS6** Evaluating, predicting, proposing

2.5 Exercise 1: Check your understanding

1. **GS2** Explain why Australia has such a wide variety of biomes.
2. **GS2** With the aid of a flow diagram, show how the Australian *environment changed* when European occupiers arrived.
3. **GS2** Explain what you understand by the term *biodiversity* and why it is important.
4. **GS2** What other types of biomes would you expect to find in Australia?
5. **GS2** Explain why most of Australia's native grasslands have disappeared.

2.5 Exercise 2: Apply your understanding

1. **GS5** Describe the *interconnection* between biomes and climate.
2. **GS6** Select one of the biomes discussed in this subtopic. Predict what might happen if it were *changed*; for example, if the wetlands were drained, or all the old-growth forests were cut down. Include a reference to the effect this would have on biodiversity.
3. **GS6** Predict what might happen if Victoria's old growth forests were logged.
4. **GS2** Explain why burning is an essential element in maintaining the grassland biome.
5. **GS2** Explain why seagrass meadows are often referred to as 'the forests of the sea'.

Try these questions in learnON for instant, corrective feedback. Go to www.jacplus.com.au.

2.6 The grassland biome

2.6.1 What are the characteristics of grasslands?

Grassland, pampas, savanna, chaparral, cerrado, **prairie**, rangeland and steppe all refer to a landscape that is dominated by grass. Once, grasslands occupied about 42 per cent of the Earth's land surface, but today they make up about 25 per cent of its land area. Grasslands are found on every continent except Antarctica.

The grassland biome is dominated by grasses, and generally has few or no trees, though there may be more tree cover in adjoining areas, such as along riverbanks. They develop in places where there is not enough rain to support a forest but there is too much rain for a desert; for this reason they are sometimes referred to as a transitional landscape.

Grasslands are found in both temperate and tropical areas where rainfall is between 250 mm and 900 mm per year. In tropical regions, grasslands tend to have a distinct wet and dry season. In temperate regions, the summers tend to be hot and the winters cool. Generally, grasslands in the southern hemisphere receive more rainfall.

Grasslands can occur naturally or as a result of human activity. The presence of large numbers of grazing animals and frequent fires prevent the growth of tree seedlings and promote the spread of grasses. Unlike other plant species, grasses can continue to grow even when they are continually grazed by animals because their growth points are low and close to the soil. Because grasses are fast-growing plants, they can support a high density of grazing animals, and they regenerate quickly after fire.

Some grasses can be up to two metres in height, with roots extending up to one metre below the soil.

FIGURE 1 Grasslands occupy about a quarter of the Earth's land surface.

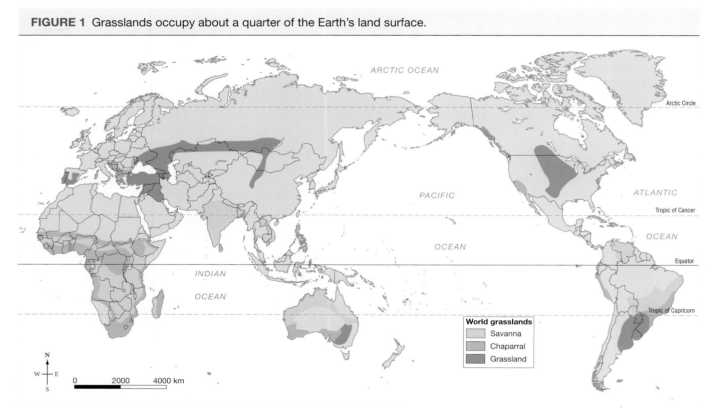

Source: Spatial Vision

2.6.2 Why are grasslands important?

Grasslands are the most useful biome for agriculture because the soils are generally deep and fertile. Almost one billion people depend on grasslands for their livelihood or as a food source. Grasslands are ideally suited for growing crops or creating pasture for grazing animals. The prairies of North America, for example, are one of the richest agricultural regions on Earth.

Grasslands are also one of the most endangered biomes and are easily turned into desert. The entire ecosystem depends on its grasses and their annual regeneration. It is almost impossible to re-establish a grassland ecosystem once desert has taken over.

Grasslands often depend on fire to germinate their seeds and generate new plant growth. Indigenous populations, such as Australian Aboriginal peoples, used this technique to flush out any wildlife that was hidden by long grass.

In more recent times, grasslands have been used for livestock grazing and are increasingly under pressure from **urbanisation**. Grasslands have also become popular tourist destinations, with people flocking to them to see majestic herds such as wildebeest, caribou and zebra, as well as the migratory birds that periodically inhabit these environments.

All the major food grains — corn, wheat, oats, barley, millet, rye and sorghum — have their origins in the grassland biome. Wild varieties of these grains are used to help keep cultivated strains disease free. Many native grass species have been used to treat diseases including HIV and cancer. Others have proven to have properties useful for treating headaches and toothache.

Grasslands are also the source of a variety of plants whose fibres can be woven into clothing. The best known and most widely used fibre is cotton. Harvested from the cottonseed, it is used to produce yarn that is then knitted or sewn to make clothing. Lesser known fibres include flax and hemp. Harvested from the stalk of the plant, both fibres are much sturdier and more rigid than cotton but can be woven to produce fabric. Hemp in particular is highly absorbent and has UV-blocking qualities.

In Australia today, less than one per cent of native grasslands survive, and they are now considered one of the most threatened Australian habitats. Since European occupation, most native grassland has been removed or changed by farming and other development. Vast areas of grassland were cleared for crops, and introduced grasses were planted for grazing animals such as sheep and cattle.

FIGURE 2 Wheat is a type of grass.

FIGURE 3 Grasslands can support a high density of grazing animals. In Australia, we use grasslands for fine wool production.

DISCUSS

Few people realise that less than one per cent of Australia's native grasslands survive. Why does such a significant loss of grassland biomes not attract the same attention as the loss of other biomes, such as our tropical rainforest and coral reefs? How would each of the following groups perceive the value of grasslands?

a. Graziers (sheep and cattle farmers)
b. City dwellers
c. Environmentalists

[Personal and Social Capability]

Explore more with my World Atlas

Deepen your understanding of this topic with related case studies and questions.
• Investigating Australian Curriculum topics > Year 9: Biomes and food security >**Wheat**

 Resources

🧩 **Interactivity** Grass, grains and grazing (int-3318)

2.6 INQUIRY ACTIVITY

Grasslands are located on six of the Earth's seven continents. Working in groups, investigate one of the grassland biomes. Using ICT, create a presentation on your chosen biome that covers the following:

a. the characteristics of the *environment*, including climate and types of grasses that dominate this *place*
b. the animals that are commonly found there
c. how the *environment* is used and *changed* for the production of food, fibre and wood products
d. threats to this particular grassland, including the *scale* of these threats
e. what is being done to manage this grassland *environment* in a *sustainable* manner.

Examining, analysing, interpreting
Classifying, organising, constructing

2.6 EXERCISES

Geographical skills key: GS1 Remembering and understanding **GS2** Describing and explaining **GS3** Comparing and contrasting **GS4** Classifying, organising, constructing **GS5** Examining, analysing, interpreting **GS6** Evaluating, predicting, proposing

2.6 Exercise 1: Check your understanding

1. **GS1** What is a grassland?
2. **GS1** Describe the global distribution of grasslands.
3. **GS1** Why are grasslands an important *environment*?
4. **GS1** Describe the major threats to this *environment*.
5. **GS2** Explain how Indigenous populations used the grassland *environment* in a *sustainable* way.
6. **GS2** Explain why so little of Australia's grasslands remain.

2.6 Exercise 2: Apply your understanding

1. **GS6** Grasslands are often referred to as a transitional landscape. Suggest a reason why grasslands might be classified in this way.
2. **GS2** Grasses are different to other plant species. Explain.
3. **GS2** Describe how grasslands differ in different climatic regions.
4. **GS2** Describe the different ways in which the grassland biome is used by people.
5. **GS6** In some *places*, attempts are being made to re-establish native grasses. Suggest why it is important to re-establish native grasslands.

Try these questions in learnON for instant, corrective feedback. Go to www.jacplus.com.au.

2.7 Coastal wetlands

2.7.1 What are wetlands?

Wetlands are biomes where the ground is saturated, either permanently or seasonally. They are found on every continent except Antarctica. Wetlands include areas that are commonly referred to as marshes, swamps and bogs. In coastal areas they are often tidal and are flooded for part of the day. In the past they were often considered a 'waste of space', and in developed nations they were sometimes drained for agriculture or the spread of urban settlements.

2.7.2 Why are wetlands important?

Wetlands are a highly productive biome. They provide important habitats and breeding grounds for a variety of marine and freshwater species. In fact, a wide variety of aquatic species that we eat, such as fish, begin their life cycle in the sheltered waters of wetlands. They are also important nesting places for a large number of migratory birds.

Wetlands are a natural filtering system and help purify water and filter out pollutants before they reach the coast. In addition, they help regulate river flow and stabilise the shoreline. **FIGURE 1** shows a cross-section through a mangrove wetland.

FIGURE 1 Cross-section of a wetland

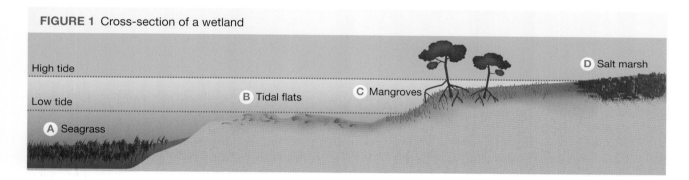

A **Seagrass meadows:**
- are covered by water all the time
- bind the mud and provide shelter for young fish
- produce **organic matter**, which is consumed by marine creatures.

B **Tidal flats:**
- are covered by tides most of the time
- are exposed for short periods of the day (low tide)
- are formed by silt and sand that has been deposited by tides and rivers
- provide a feeding area for birds and fish.

C **Mangroves:**
- have **pneumatophores** that trap sediment and pollutants from the land and sea
- change shallow water into swampland
- store water and release it slowly into the ecosystem
- have leaves that decompose and provide a food source for marine life
- provide shelter, breeding grounds and a nursery for marine creatures and birds.

D **Salt marshes:**
- are covered by water several times per year
- provide decomposing plant matter — an additional food source for marine life
- have high concentrations of salt.

Dalywoi Bay in the Northern Territory is a coastal mangrove wetlands area. The topographic map in **FIGURE 2** shows the features of this landscape.

FIGURE 2 Dalywoi Bay, Northern Territory

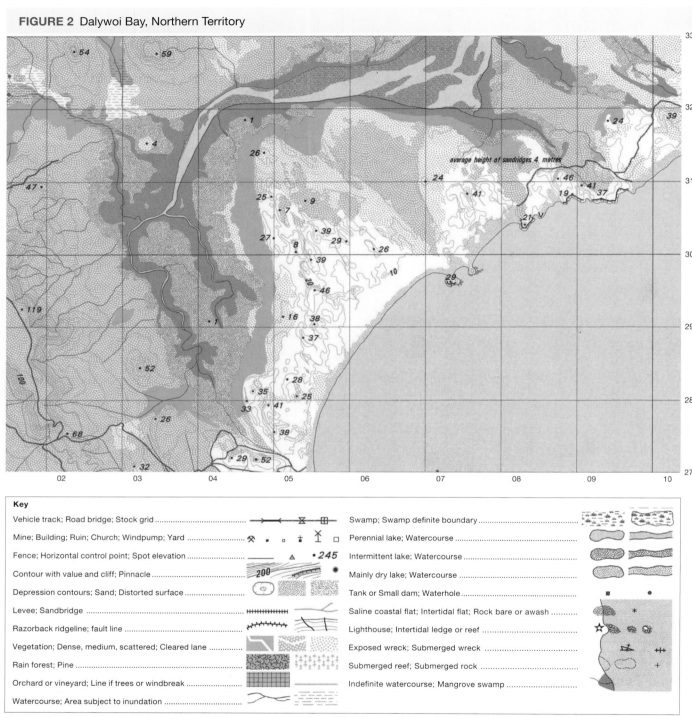

Key

Vehicle track; Road bridge; Stock grid	Swamp; Swamp definite boundary
Mine; Building; Ruin; Church; Windpump; Yard	Perennial lake; Watercourse
Fence; Horizontal control point; Spot elevation	Intermittent lake; Watercourse
Contour with value and cliff; Pinnacle	Mainly dry lake; Watercourse
Depression contours; Sand; Distorted surface	Tank or Small dam; Waterhole
Levee; Sandbridge	Saline coastal flat; Intertidal flat; Rock bare or awash
Razorback ridgeline; fault line	Lighthouse; Intertidal ledge or reef
Vegetation; Dense, medium, scattered; Cleared lane	Exposed wreck; Submerged wreck
Rain forest; Pine	Submerged reef; Submerged rock
Orchard or vineyard; Line if trees or windbreak	Indefinite watercourse; Mangrove swamp
Watercourse; Area subject to inundation	

Source: The Australian Army © Commonwealth of Australia 1999

FIGURE 3 Seagrass

FIGURE 4 Pneumatophores — the exposed root system of mangroves that enables them to take in air when the tide is in

2.7 EXERCISES

Geographical skills key: GS1 Remembering and understanding **GS2** Describing and explaining **GS3** Comparing and contrasting **GS4** Classifying, organising, constructing **GS5** Examining, analysing, interpreting **GS6** Evaluating, predicting, proposing

2.7 Exercise 1: Check your understanding

1. **GS2** How have mangroves adapted to survive in their marine–terrestrial *environment*?
2. **GS2** What are seagrass meadows and why are they important?
3. **GS1** Define the term 'wetland'.
4. **GS2** Coastal wetlands are tidal. What does this mean?
5. **GS6** Wetlands were once described as 'a waste of *space*'. Do you think this is an accurate description? Give reasons for your answer.

2.7 Exercise 2: Apply your understanding

1. **GS5** Refer to **FIGURE 2** and describe the landscape at the following grid references: 042309, 071329, 030320 and 055290.
2. **GS6** In **FIGURE 2**, locate the grid square inside grid references: 030300, 030310, 040030, 040310.
   (a) What is the area covered by these grid squares?
   (b) Describe how this *environment* would *change* over the course of the day.
   (c) A proposal has been put forward to construct a **canal housing estate** in this location. It is proposed that the estate will occupy this grid square and its surrounds. Describe the *scale* of this project.
   (d) Explain how the *environment* may *change* if this project goes ahead.
3. **GS3** Distinguish between a tidal flat and a salt marsh.
4. **GS5** Wetlands have been described as a natural purification system. Which part of the wetland *environment* would perform this function? Give reasons for your answer.
5. **GS6** Describe what might happen if there were no wetlands.

Try these questions in learnON for instant, corrective feedback. Go to www.jacplus.com.au.

2.8 SkillBuilder: Constructing and describing a transect on a topographic map

What is a transect?

A transect is a cross-section with additional detail, which summarises information about the environment. In addition to the shape of the land, a transect shows what is on the ground, including landforms, vegetation, soil types, settlements and infrastructure.

Select your learnON format to access:

- an explanation of the skill (Tell me)
- a step-by-step process to develop the skill, with an example (Show me)
- an activity to allow you to practise the skill (Let me do it)
- questions to consolidate your understanding of the skill.

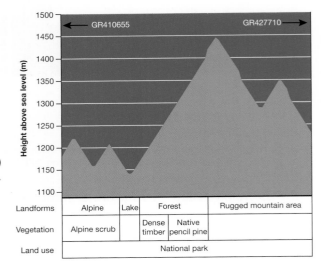

---

on Resources

Video eLesson Constructing and describing a transect on a topographic map (eles-1727)

Interactivity Constructing and describing a transect on a topographic map (int-3345)

2.9 Coral reefs

2.9.1 Formation of coral reefs

Coral reefs are found in spaces around tropical and subtropical shores. They require specific temperatures and sea conditions and an area free from sediment. Coral reefs are built by tiny animals called **coral polyps**. The upper layer is alive, growing on the remains of millions of dead coral. Coral reefs are one of the oldest ecosystems on Earth and also one of the most vulnerable to human activity.

Coral reefs are one of the most diverse biomes on Earth and are built by polyps that live in groups. A reef is a layer of living coral growing on the remains of millions of layers of dead coral. There are inner and outer reefs as well as coral cays (small islands of coral) and coral atolls.

FIGURE 1 outlines the anatomy of a coral reef, and provides a close-up image of some of the millions of coral polyps that combine to form a reef. **FIGURE 2** shows how different reefs are formed over time.

FIGURE 1 Anatomy of a coral reef

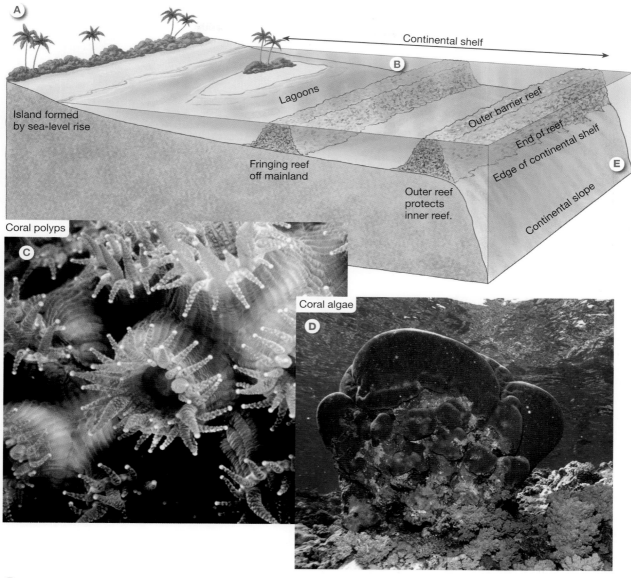

Continental shelf

Lagoons

Island formed
by sea-level rise

Outer barrier reef

End of reef

Edge of continental shelf

Fringing reef
off mainland

Outer reef
protects
inner reef.

Continental slope

Coral polyps

Coral algae

(A) Continental island and fringing reef

(B) • Corals form in warm shallow saltwater where the temperature is between 18 °C and 26 °C.
  • Water must be clear, with abundant sunlight and gentle wave action to provide oxygen and distribute nutrients.

(C) Coral polyps have soft, hollow bodies shaped like a sac with tentacles around the opening. They cover themselves in a limestone skeleton and divide and form new polyps.

(D) Producers such as algae give coral its colour and provide a food source for marine life such as fish. Coral reefs support at least one-third of all marine species. They are the marine equivalent of the tropical rainforest.

(E) Beyond the continental shelf, the water is too deep and cold for coral. Sunlight cannot penetrate to allow coral growth.

1. Fringing coral reefs develop along the shores of continents and islands.

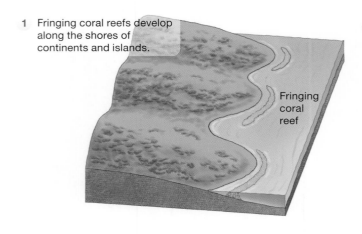

Fringing coral reef

2. When sea levels rise, fringing reefs become barrier reefs.

Barrier reef

3. Formation of a coral atoll

a. Volcanic island

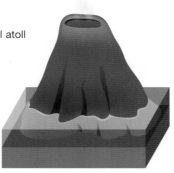

b. Eroded volcanic island has been partly submerged by rising sea

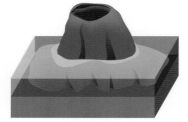

c. Sandy island forms on reef from eroded coral and shell

What was once an island is now completely submerged.

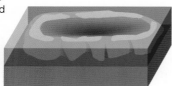

2.9.2 Benefits of coral reefs

Today, about 500 million people rely on reef systems, either for their livelihood, as a source of food, or as a means of protecting their homes along the coastline. (Coral reefs help break up wave action, so waves have less energy when they reach the shoreline, thus reducing coastal erosion.) It is estimated that coral reefs contribute between US$28.8 billion and US$375 billion to the global economy each year.

Reefs are important to both the fishing and tourism industries. Approximately 2 million people visit Australia's Great Barrier Reef Marine Park alone, generating more than A$2 billion for the local economy. Nearly one-third of all tourists who visit Australia also visit the Great Barrier Reef.

Coral reefs have been found to contain compounds vital to the development of new medicines.
- Painkillers have been developed from the venom of cone shells.
- Some cancer treatments come from algae.
- Treatments for cardiovascular disease and HIV include compounds that were originally found in coral reefs.

2.9.3 Threats to coral reefs

Reefs face a variety of threats.

- Urban development requires land clearing and wetland drainage, which increases erosion. Sediment washed into water prevents sunlight penetrating the water.
- Contamination by fossil fuels, chemical waste and agricultural fertilisers pollutes the sea.
- Tourism damages coral through boats dropping anchor, or tourists taking coral or walking on it.
- Global warming increases water temperature, which bleaches the coral, turning it white and destroying the reef system.
- Predators, such as the crown of thorns starfish, prey on coral polyps, which affects the whole ecosystem.

Explore more with my**World**Atlas

Deepen your understanding of this topic with related case studies and questions.
- Investigate additional topics > Environments > **The Great Barrier Reef**

2.9 INQUIRY ACTIVITIES

1. With the aid of a diagram, explain how coral reefs are formed. **Describing and explaining**
2. Investigate two of the threats to coral reefs and prepare an annotated visual display that outlines:
   (a) the nature of the threat
   (b) the **changes** that will occur or have occurred as a result of this threat
   (c) the impact of these **changes** on the **environment**, including references to the rate and **scale** of the **changes**
   (d) a strategy for the long-term **sustainable** management of the reef **environment**.

Examining, analysing, interpreting
Classifying, organising, constructing

2.9 EXERCISES

Geographical skills key: GS1 Remembering and understanding **GS2** Describing and explaining **GS3** Comparing and contrasting **GS4** Classifying, organising, constructing **GS5** Examining, analysing, interpreting **GS6** Evaluating, predicting, proposing

2.9 Exercise 1: Check your understanding

1. **GS2** Explain the difference between a fringing reef and a barrier reef.
2. **GS2** Detail the steps in the formation of a coral reef.
3. **GS2** Why are coral reefs important **places**?
4. **GS1** What gives coral its colour?
5. **GS1** Outline five key threats to coral reefs.

2.9 Exercise 2: Apply your understanding

1. **GS6** Coral reefs are highly susceptible to climate **change**.
   (a) Explain what you understand by the term **climate change**.
   (b) Explain how the coral reef **environment** would **change** if sea temperatures were to rise by 2 °C.
2. **GS3** Coral reefs and rainforests have a lot in common. With the aid of a Venn diagram, show the similarities and differences between these two biomes.
3. **GS5** The English naturalist, biologist and evolutionary scientist Charles Darwin once described coral reefs as being like an oasis in the desert. Explain what you think Darwin meant by this analogy.
4. **GS2** Describe the ideal growing conditions for coral.
5. **GS2** Coral reefs have a 'garden-like' appearance, but they are not plants. Explain.

Try these questions in learnON for instant, corrective feedback. Go to www.jacplus.com.au.

2.10 Managing and protecting biomes

2.10.1 Can we recreate what has been lost?

It is important to consider the long-term impact of our actions and ensure that they do not have a negative impact on the environment. In the past, human activity has polluted and degraded the land. However, we are now attempting to learn from our mistakes and repair the damage done.

The role of Biosphere 2

Biosphere 2 (see **FIGURES 1** and **2**) is a research facility in the Arizona Desert in the USA. It has been designed to investigate Earth's life systems. Covering an area of 1.5 hectares and standing almost 28 metres at its highest point, Biosphere 2 is the world's largest ecological laboratory. Its mission is to learn more about how the environment connects with us and how we in turn connect with the environment.

FIGURE 1 Aerial view of Biosphere 2

Within the confines of Biosphere 2, five natural environments have been recreated: rainforest, desert, savanna, wetland and ocean (including a coral reef). In addition, agricultural and human living spaces have also been created. More than 3000 living organisms are found across the complex. All systems, including oxygen levels, water supply and climate are managed by complex computer systems within the facility.

While early experiments focused on our ability to artificially recreate the Earth's biosphere and sustain life within it, the emphasis has now moved to investigating the impact of human activity. Scientists are looking at how increased burning of fossil fuels and the destruction of habitats will affect the natural systems that sustain all life on the planet.

The project has helped demonstrate the complexity of the natural processes that occur on Earth and within particular biomes. Constant work, effort and thought is needed to maintain the natural order.

FIGURE 2 Ocean recreated within Biosphere 2

2.10.2 Landcare and other action groups

Landcare was born in 1986, when a group of farmers near St Arnaud in central Victoria banded together to find sustainable solutions to their common problem: **land degradation**. The idea has since been adopted by the Australian government and has spread nationwide. Landcare is about communities working together on environmental projects such as:

- cleaning up polluted creeks and waterways
- planting trees
- restoring beach dune systems
- finding workable solutions to problems such as **salinity** in farming communities
- addressing the growing problem of waste disposal and plastic bags.

Other organisations such as Clean Up Australia and OceanCare operate with a similar vision of protecting our waterways and oceans from the impacts of land degradation and pollution.

FIGURE 3 Members of the Tangaroa Blue OceanCare group with some of the rubbish collected during a beach clean-up day

2.10.3 Learning from Indigenous communities

Long before the arrival of European colonisers, Indigenous Australian peoples practised their own form of agriculture. Rather than simply hunting and gathering, they used knowledge amassed over thousands of years to manage the native plants and animals they relied on for sustenance. Fire formed the basis of the Indigenous land management system. A complex system of burning ensured that food supplies were both sustainable and predictable.

Careful planning enabled them to change the landscape to meet their needs. First they created a grassland devoid of trees, the food source for grazing animals. Adjacent to this they cleared out the undergrowth and thinned the trees to create an open forest area that would provide shelter for these same grazing animals. Then they burned the grassland to create the new growth that would encourage the grazers from the shelter of the trees and make hunting easier.

Burning was usually carried out at night time or in the early morning to produce a 'cool fire'. The evening dew helped control the heat produced and made the fire easier to control. Such fires were also self-extinguishing; once the grass was burned, the fire simply went out.

This was preferable to a 'hot fire' fanned by the flammable oils sweated by plants in the heat of the day, which could easily get out of control.

Indigenous Protected Areas

Indigenous Protected Areas (IPAs) are wholly managed using traditional practices. Generational knowledge is passed down and preserved for future generations. Approximately 36 per cent of Australia's protected areas is under the control of local Indigenous communities. The first IPA was established at Nantawarrina, about 555 kilometres north of Adelaide, in 1998. The largest is Southern Tanami in the Northern Territory, at 10.16 million hectares, and the smallest is Pulu Islet in the Torres Strait, at about 15 hectares.

FIGURE 4 Map of Indigenous Protected Areas

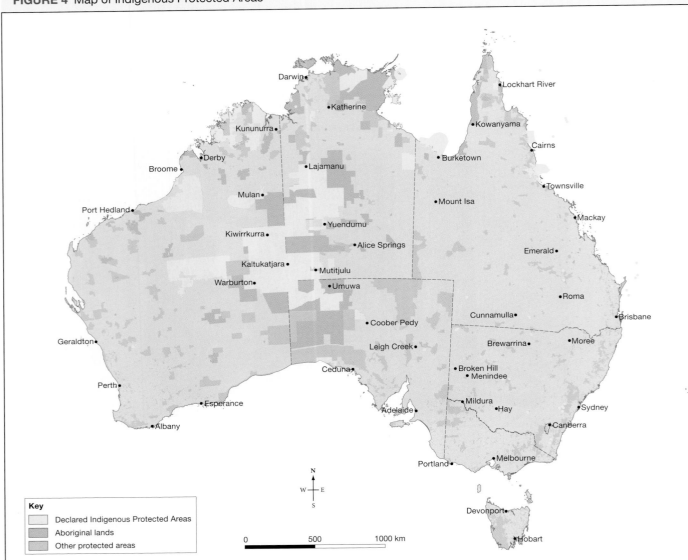

Source: © Commonwealth of Australia 2013, Department of Environment

---

2.10 INQUIRY ACTIVITIES

1. Use the internet to learn more about Biosphere 2 and a similar facility, the Eden Project, in England. Display your findings as an annotated visual display.
   Examining, analysing, interpreting
   Classifying, organising, constructing

2. Investigate what projects have been carried out in your local area to restore the natural **environment.**
   Examining, analysing, interpreting

2.10 EXERCISES

Geographical skills key: GS1 Remembering and understanding **GS2** Describing and explaining **GS3** Comparing and contrasting **GS4** Classifying, organising, constructing **GS5** Examining, analysing, interpreting **GS6** Evaluating, predicting, proposing

2.10 Exercise 1: Check your understanding

1. **GS1** How did Indigenous land management differ from that of European colonisers?
2. **GS1** What is Biosphere 2?
3. **GS2** Explain the purpose of Landcare.
4. **GS1** Identify three different types of projects that Landcare has undertaken.
5. **GS1** Why was the use of fire an integral part of land management for Indigenous communities?

2.10 Exercise 2: Apply your understanding

1. **GS6** Do you think projects such as Biosphere 2 serve a useful purpose? Give reasons for your answer.
2. **GS6** The media is constantly filled with information about the state of the **environment**. Write a letter to the editor in response to the following statement: 'Biosphere 2 has done little to increase or improve our understanding of the **environment**, which continues to deteriorate. Our energies should be directed towards the **sustainable** use of the resources we have, rather than being frittered away on such experiments, which are nothing more than expensive toys.'
3. **GS6** Why do you think Indigenous Protected Areas have been established?
4. **GS5** Each year bushfires cause an enormous amount of damage to biomes.
   (a) Why were uncontrolled fires relatively unknown prior to the arrival of European colonisers?
   (b) Do you think more widespread use of Indigenous land management methods could prevent such widespread destruction? Justify your point of view.
5. **GS2** Consider Indigenous land management techniques.
   (a) Outline the difference between a 'cool fire' and a 'hot fire'.
   (b) Explain why a 'cool burn' is used as a form of land management.

Try these questions in learnON for instant, corrective feedback. Go to www.jacplus.com.au.

2.11 Thinking Big research project: Our world of biomes AVD

SCENARIO

The Department of the Environment has commissioned you to carry out an in-depth study of biomes, their characteristics and the factors that influence their development. You will create an engaging annotated visual display to showcase your findings.

Select your learnON format to access:

- the full project scenario
- details of the project task
- resources to guide your project work
- an assessment rubric.

 Resources

ProjectsPLUS Thinking Big research project: Our world of biomes AVD (pro-0188)

2.12 Review

2.12.1 Key knowledge summary
Use this dot point summary to review the content covered in this topic.

2.12.2 Reflection
Reflect on your learning using the activities and resources provided.

 Resources

 eWorkbook Reflection (doc-31714)

Crossword (doc-31715)

 Interactivity All the world is a biome crossword (int-7644)

KEY TERMS

biodiversity the variety of plant and animal life within an area

canal housing estate a housing estate built upon a system of waterways, often as the result of draining wetland areas. All properties have water access.

clear-felling the removal of all trees in an area

coral polyp a tube-shaped marine animal that lives in a colony and produces a stony skeleton. Polyps are the living part of a coral reef.

deforestation clearing forests to make way for housing or agricultural development

land degradation deterioration in the quality of land resources caused by excessive exploitation

latitude the angular distance north or south from the equator of a point on the Earth's surface

leaching the process by which water runs through soil, dissolving minerals and carrying them into the subsoil

leeward describes the area behind a mountain range, away from the moist prevailing winds

logging large-scale cutting down, processing and removal of trees from an area

organic matter decomposing remains of plant or animal matter

pneumatophores exposed root system of mangroves, which enables them to take in air when the tide is in

prairie native grassland of North America

precipitation the forms in which moisture is returned to the Earth from the sky, most commonly in the form of rain, hail, sleet and snow

rain shadow the dry area on the leeward side of a mountain range

salinity the presence of salt on the surface of the land, in soil or rocks, or dissolved in rivers and groundwater

treeline the edge of the area in which trees are able to grow

tundra the area lying beyond the treeline in polar or alpine regions

urbanisation the growth and spread of cities

windward describes the side of the mountain that faces the prevailing winds

3 Feeding the world

3.1 Overview

Everyone needs to eat. How does the world produce all the food it needs, and is there a better way?

3.1.1 Introduction

Food is a fundamental part of every person's life. For many people, what to eat each day can be a constant thought and for some, a constant worry. What are the foods we eat, and why do these vary across the globe? How do we modify biomes to produce the food we need, and how can we build on our understanding of food sourcing and production strategies to feed the world's future generations?

To access a pre-test and starter questions and receive immediate, **corrective feedback** and **sample responses** to every question, select your learnON format at www.jacplus.com.au.

3.2 What does the world eat?

3.2.1 The major food staples

Staple foods are those that are eaten regularly and in such quantities that they constitute a dominant portion of a diet. They form part of the normal, everyday meals of the people living in a particular place or country. Staple foods vary from place to place, but are typically inexpensive or readily available. The staple food of an area is normally interconnected to the climate of that area and the type of land.

Most staple foods are cereals, such as wheat, barley, rye, oats, maize (corn) and rice; or root vegetables, such as potatoes, yams, taro and cassava. Rice, maize and wheat provide 60 per cent of the world's food energy intake; 4 billion people rely on them as their staple food.

Other staple foods include legumes, such as soya beans and sago; fruits, such as breadfruit and plantains (a type of banana); and fish.

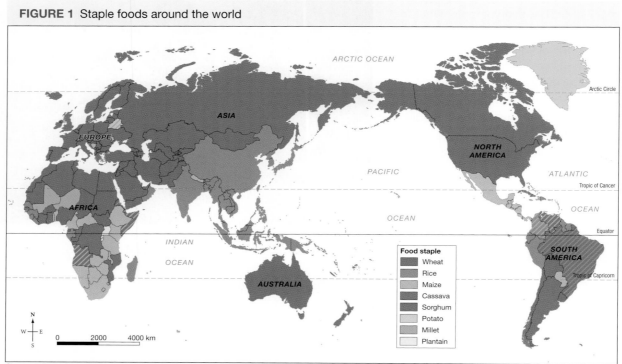

FIGURE 1 Staple foods around the world

Source: Data from FAO

3.2.2 Wheat, maize and fish

Wheat

Wheat is a cereal grain that is cultivated across the world. In 2019, the total world production of wheat was nearly 735 million tons, making it the second most produced cereal after maize (1.1 billion tons) and above rice (496 million tons). World trade in wheat is greater than for all other crops combined.

Wheat was one of the first crops to be easily cultivated on a large scale, with the added advantage of yielding a harvest that could be stored for a long time. Wheat covers more land area than any other commercial crop and is the most important staple food for humans.

FIGURE 2 Wheat is used in a wide variety of foods such as breads, biscuits, cakes, breakfast cereals and pasta.

Maize

Maize, or corn, was commonly grown throughout the Americas in the late fifteenth and early sixteenth centuries. Explorers and traders carried maize back to Europe and introduced it to other countries. It then spread to the rest of the world, thanks to its ability to grow in different environments. Sugar-rich varieties called sweet corn are usually grown for human consumption, while field corn varieties are used for animal feed and **biofuel**. Maize is the most widely grown grain crop in the Americas, covering 70–100 million acres of farmland in the US alone, which accounts for nearly 40 per cent of all maize grown in the world.

TABLE 1 Top 10 maize producers, 2019

Country	Production (million tonnes)
United States	377.5
China	224.9
Brazil	83.0
India	42.3
Argentina	40.0
Ukraine	39.2
Mexico	32.6
Indonesia	20.8
France	17.1
South Africa	15.5

Source: www.thedailyrecords.com

FIGURE 3 Corn cobs drying outside in Serbia

Fish

Fish is a staple food in many societies. The oceans provide an irreplaceable, renewable source of food and nutrition essential to good health. In general, people in developing countries, especially those in coastal areas, are much more dependent on fish as a staple food than those in the developed world. About 3 billion people rely on fish as their primary source of animal protein.

FIGURE 4 A fish haul in Bali, Indonesia

Explore more with myWorldAtlas

Deepen your understanding of this topic with related case studies and questions.
- Investigating Australian Curriculum topics > Year 9: Biomes and food security > **Wheat**

3.3 How can we feed the world?

3.3.1 Challenges to feeding the global population

At the beginning of the twentieth century, the entire world population was less than 2 billion people. The current world population is more than 7.7 billion. Earth's population is projected to rise to 9 billion people by 2050, and we all need food. What can we do to ensure there is enough food for everyone?

FIGURE 1 shows that crops occupy half the available agricultural land space. Almost all future population growth will occur in the developing world. This increased population, combined with higher standards of living in developing countries, will create enormous strains on land, water, energy and other natural resources.

There is currently about one-sixth of a hectare of **arable** land **per capita** in East and South Asia. With population growth, and almost no additional land available for agricultural expansion, arable land per capita will continue to decline.

3.3.2 Food production increases

Agricultural yields vary widely around the world depending on climate, management practices and the types of crops grown. Globally, 15 million square kilometres of land are used for growing crops — altogether, that's about the size of South America. Approximately 32 million square kilometres of land around the world are used for pasture — an area about the size of Africa. Across the Earth, most land that is suitable for agriculture is already used for that purpose and, in the past 50 years, we have increased our food production.

According to the UN Food and Agriculture Organization (FAO), the three main factors that have affected recent increases in world crop food production are:

- increased cropland and rangeland area
- increased yield per unit area
- greater cropping intensity.

FIGURE 1 World distribution of cropland, pasture and maize. More maize could be grown if improvements were made to seeds, irrigation, fertiliser and markets.

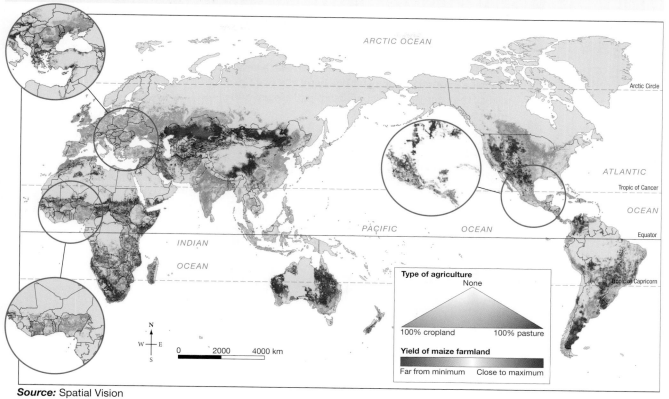

Source: Spatial Vision

FAO projections suggest that cereal demand will increase by almost 50 per cent by 2050. This can either be obtained by increasing yields, expanding cropland through conversion of natural habitats, or growing crops more efficiently. **FIGURE 2** shows the growth in crop yields in developing countries from 1961 to what is predicted for 2030. Rice, maize and wheat have had significant increases in yield.

FIGURE 2 Crop yields in developing countries, 1961–2030

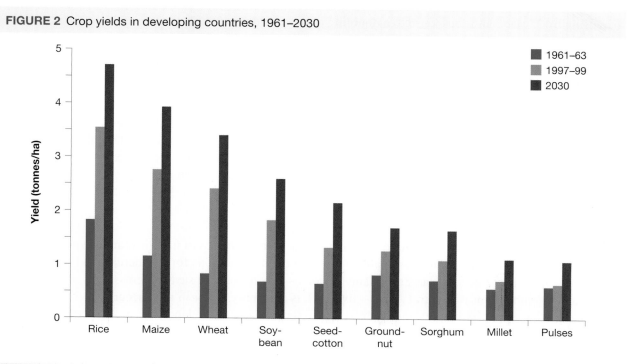

Agricultural **innovations** have also changed and increased global food production. They have boosted crop yields through advanced seed genetics, agronomic practices (scientific production of food plants) and product innovations that help farmers maximise productivity and quality (see **FIGURE 3**). In this way, the nutritional content of crops can be increased.

FIGURE 3 Farmers in a village in Kenya use a laptop to examine information on plant diseases at a plant health clinic. They can also consult a plant pathologist and show them samples of their crops.

3.3.3 Increasing our food production

World food production has grown substantially over the past century. Increased fertiliser application and more water usage through irrigation have been responsible for over 70 per cent of crop yield increases. The Second Agricultural Revolution in developed countries after World War II and the **Green Revolution** in developing countries in the mid 1960s transformed agricultural practices and raised crop yields dramatically.

Since the 1960s agriculture has been more productive, with world per capita agricultural production increasing by 25 per cent in response to a doubling of the world population.

It is possible to get even more food out of the land we are already using. For example, **FIGURE 1** shows the places where maize yields could increase and become more **sustainable** by improving nutrient and water management, seed types and markets.

Environmental factors

In the past, growth in food production resulted mainly from increased crop yields per unit of land and to a lesser extent from expansion of cropland. From the early 1960s until 2014, total world cropland increased by only around 10 per cent, but total agricultural production grew by 60 per cent. Increases in yields of crops, such as sweet potatoes and cereals, were brought about by a combination of:
- increased agricultural inputs
- more intensive use of land
- the spread of improved crop varieties.

In some places, such as parts of Africa and South-East Asia, increases in fisheries (areas where boats are used to catch fish) and expansion of cropland areas were the main reasons for the increase in food supply. In addition, cattle herds became larger. In many regions — such as in the savanna grasslands of Africa, the Andes, and the mountains of Central Asia — livestock is a primary factor in food security today. Fertilisers have increased agricultural outputs and enabled more intensive use of the land. The global fertiliser use of 208 million tonnes in 2020 represents a 30 per cent increase since 2008.

TABLE 1 Fertiliser use, 1959–60, 1989–90 and 2020

Region/nutrient	Fertiliser use			Annual growth	
	1959–60	1989–90	2020	1960–90	1990–2020
	(million nutrient tonnes)			(per cent)	
Developed countries	24.7	81.3	86.4	4.0	0.2
Developing countries	2.7	62.3	121.6	10.5	2.2
East Asia	1.2	31.4	55.7	10.9	1.9
South Asia	0.4	14.8	33.8	12.0	2.8
West Asia/North Africa	0.3	6.7	11.7	10.4	1.9
Latin America	0.7	8.2	16.2	8.2	2.3
Sub-Saharan Africa	0.1	1.2	4.2	8.3	3.3
World total	27.4	143.6	208.0	5.5	1.2
Nitrogen	9.5	79.2	115.3	7.1	1.3
Phosphate	9.7	37.5	56.0	4.5	1.3
Potash	8.1	26.9	36.7	4.0	1.0

Source: Bumb, B. and C. Baanante. 1996. World Trends in Fertilizer Use and Projections to 2020. Policy Brief 38, Table 1. Washington, DC: International Food Policy Research Institute http://www.ifpri.org/publication/world-trends-fertilizer-use-and-projections-2020

Trade factors and economic factors

From the 1960s onwards, there has been significant growth of world trade in food and agriculture. Food and fertiliser imports by developing countries have grown, reducing the threat of famine in those countries.

TABLE 2 Percentage share of crop production increases, 1961–2030

	Arable land expansion (1)		Increases in cropping intensity (2)		Harvested land expansion (1+2)		Yield increases	
	1961–99	1997/99 –2030	1961–99	1997/99 –2030	1961–99	1997/99 –2030	1961–99	1997/99 –2030
All developing countries	23	21	6	12	29	33	71	67
South Asia	6	6	14	13	20	19	80	81
East Asia	26	5	−5	14	21	19	79	81
East/North Africa	14	13	14	19	28	32	72	68
Latin America and the Caribbean	46	33	−1	21	45	54	55	46
Sub-Saharan Africa	35	27	31	12	66	39	34	61
World	15		7		22		78	

3.3.4 The impact of the Green Revolution

The Green Revolution was a result of the development and planting of new **hybrids** of rice and wheat, which led to greatly increased yields. There have been a number of green revolutions since the 1950s, including those in:

- the United States, Europe and Australia in the 1950s and 1960s
- New Zealand, Mexico and many Asian countries in the late 1960s, 1970s and 1980s.

The Green Revolution saw a rapid increase in the output of cereal crops — the main source of calories in developing countries. Farmers in Asia and Latin America widely adopted high-yielding varieties. Governments, especially those in Asia, introduced policies that supported agricultural development. In the 2000s, cereal harvests in developing countries were triple those of 40 years earlier, while the population was only a little over twice as large.

Planting of high-yield crop varieties coincided with expanded irrigation areas and fertiliser use, leading to significant increases in cereal output and calorie availability.

FIGURE 4 Applying fertiliser to crops in the Punjab, India

3.3 INQUIRY ACTIVITIES

1. **FIGURE 1** shows where more crops could be grown. Investigate how Mexico or a country in West Africa or Eastern Europe could improve the **sustainability** of its agriculture. Create a mind map or flow chart diagram to represent your findings. **Classifying, organising, constructing**
2. Research the background of the Green Revolution — why it occurred, the key **places** involved and the **changes** that resulted. Create a dot-point summary of your findings. **Examining, analysing, interpreting**
3. Some scientists are suggesting that there will be a new Green Revolution. Investigate current thinking and predict the potential **scale** of this possible agricultural **change**. **Examining, analysing, interpreting** **Evaluating, predicting, proposing**

3.3 EXERCISES

Geographical skills key: GS1 Remembering and understanding **GS2** Describing and explaining **GS3** Comparing and contrasting **GS4** Classifying, organising, constructing **GS5** Examining, analysing, interpreting **GS6** Evaluating, predicting, proposing

3.3 Exercise 1: Check your understanding

1. **GS1** Refer to **FIGURE 1** and describe the distribution of **places** in the world with pasture and grasslands.
2. **GS1** How could crop production be increased in **places** such as Eastern Europe or Western Africa?
3. **GS2** Explain the impact of an increasing population on world **environments**.
4. **GS2** Explain why agricultural innovations can **change** food production.
5. **GS1** In the past, what were the two reasons for the increase in food production?
6. **GS1** Refer to **TABLE 1**. Describe the trends in the use of fertilisers from 1960 to 2020.
7. **GS2** Explain the significance of trade in food production.
8. **GS2** Discuss the three reasons for improved crop production.

3.3 Exercise 2: Apply your understanding

1. **GS6** Consider **FIGURE 1**. Suggest reasons why some regions are much higher crop producers than others.
2. **GS2** Were the **changes** brought about during the Green Revolution successful? Explain.
3. **GS6** With reference to specific **places**, suggest how increasing population densities might influence future crop production.

3.4 Modifying biomes for agriculture

3.4.1 Using technology for food production

In the twentieth century, rapid global population growth gave rise to serious concerns about the ability of agriculture to feed humanity. However, newer processes and technology produced additional gains in food production.

Across the world, people have modified biomes to produce food through the application of innovative technologies. In general, the focus of agriculture is to modify water, climate, soils, land and crops.

3.4.2 How do we modify climate?

Irrigation is the artificial application of water to the land or soil to supplement natural rainfall. It helps to increase agricultural production in dry areas and during periods of inadequate rainfall. (See sections 3.5.3 and 3.8.3 for examples of irrigation farming.)

In flood irrigation, water is applied and distributed over the soil surface by gravity. It is by far the most common form of irrigation throughout the world, and has been practised in many areas, virtually unchanged, for thousands of years.

Modern irrigation methods include computer-controlled drip systems that deliver precise amounts of water to a plant's root zone.

Another way of modifying climate is with the use of greenhouses (or glasshouses), which are used for growing flowers, vegetables, fruits and tobacco (see **FIGURE 1**). Greenhouses provide an artificial biotic environment to protect crops from heat and cold and to keep out pests. Light and temperature control allows greenhouses to turn non-arable land into arable land, thereby improving food production in marginal environments. Greenhouses allow crops to be grown throughout the year, so they are especially important in high-latitude countries.

FIGURE 1 Vegetables growing inside a greenhouse

The largest expanse of plastic greenhouses in the world is around the city of Almeria, in south-east Spain (see **FIGURE 2**). Here, since the 1970s, semi-arid pasture land has been replaced by greenhouse **horticulture**. Today, Almeria has become Europe's market garden. In order to grow food all year round, the region has about 26 000 hectares of greenhouses.

FIGURE 2 False-colour satellite image of greenhouses in the Almeria region

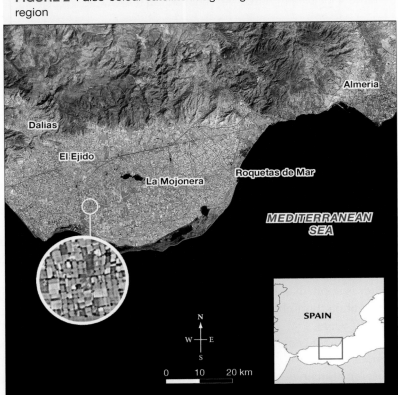

Source: American Geophysical Union

3.4.3 How do we modify soils?

Fertilisers are organic or inorganic materials that are added to soils to supply one or more essential plant nutrients. As discussed in subtopic 3.3, fertilisers play a key role in producing high-yield harvests; it is estimated that about 40 to 60 per cent of crop yields are due to fertiliser use, and that by adding fertiliser to crops, food for almost half the people on Earth is produced.

3.4.4 How do we modify landscapes?

People change landscapes in order to produce food. **Undulating** land can be flattened, steep slopes terraced, or stepped, and wetlands drained. Land reclamation is the process of creating new land from seas, rivers or lakes. In addition, it can involve turning previously unfarmed land, or degraded land, into arable land by fixing major deficiencies in the soil's structure, drainage or fertility.

In the Netherlands, the Dutch have tackled huge reclamation schemes to add land area to their country. One such scheme is the IJsselmeer (see **FIGURE 3**), where four large areas (*polders*) have been reclaimed from the sea, adding an extra 1650 square kilometres for cultivation. This has increased the food supply in the Netherlands and created an overspill town for Amsterdam.

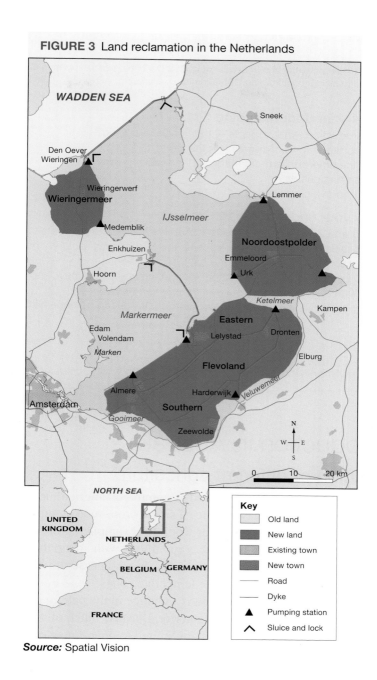

FIGURE 3 Land reclamation in the Netherlands

Source: Spatial Vision

 **Resources**

Interactivity Changing nature (int-3321)

Geographical skills key: GS1 Remembering and understanding **GS2** Describing and explaining **GS3** Comparing and contrasting **GS4** Classifying, organising, constructing **GS5** Examining, analysing, interpreting **GS6** Evaluating, predicting, proposing

3.4 Exercise 1: Check your understanding

1. **GS1** To improve food production, what aspects of biomes are modified?
2. **GS1** What is the most common form of irrigation in the world?
3. **GS1** What *changes* to the *environment* are made by land reclamation?
4. **GS1** What is greenhouse horticulture?
5. **GS2** How do fertilisers improve crop yields?

3.4 Exercise 2: Apply your understanding

1. **GS6** Refer to **FIGURES 1** and **2**. How do greenhouses modify *spaces* and *places* on the Earth's surface?
2. **GS2** How is land that is reclaimed from the sea, such as the Netherlands' *polders*, made productive for farming and food production?
3. **GS2** Refer to **FIGURE 3**. Use the *scale* to calculate the approximate area of new land created in Flevoland.
4. **GS6** Study **FIGURE 3**. What might be the purpose of the pumping stations?
5. **GS2** People can modify landscapes in order to produce food. What can be done with:
   (a) undulating land
   (b) steep slopes
   (c) wetlands?

Try these questions in learnON for instant, corrective feedback. Go to www.jacplus.com.au.

3.5 Food production in Australia

3.5.1 The distribution of different agriculture

Modern food production in Australia can be described as commercial agricultural practices that produce food for local and global markets. Farms may produce a single crop, such as sugar cane, or they may be mixed farms that produce cereal grains and sheep for wool, for example. Farms today use sophisticated technology, and in many cases are managed by large corporations with an **agribusiness** approach.

There is a wide range of agriculture types in Australia, as shown in **FIGURE 1**. They occupy space across all biomes found in Australia, from the tropics to the temperate zones.

The location of farms in Australia shows that there is a change in the pattern of farming types, from the well-watered urban coastal regions towards the arid interior. Because much of Australia's inland rainfall is less than 250 millimetres, farm types in these places are limited to open-range cattle and sheep farming.

The pattern of land use and transition of farm types is shown in **FIGURE 2**. It illustrates that **intensive farms**, which produce perishables such as fruit and vegetables, are located on high-cost land close to urban markets. At the other extreme, the **extensive farms**, which manage cattle for meat and sheep for wool, are found on the less expensive lands distant from the market.

3.5.2 Types of farming in Australia

Extensive farming of sheep or cattle

Sometimes known as livestock farming or grazing, sheep and cattle stations are found in semi-arid and desert grassland biomes, with rainfall of less than 250 millimetres. In 2017 Australia's 26 million cattle were predominantly farmed in Queensland, New South Wales and Victoria, while our 72 million sheep were found mainly in New South Wales, Victoria and Western Australia. Farms are generally large in scale, sometimes covering hundreds of square kilometres. These days, they have very few employees, and often use helicopters and motor vehicles for mustering. Meat and wool products go to both local and overseas markets for cash returns.

FIGURE 1 Types of agriculture in Australia

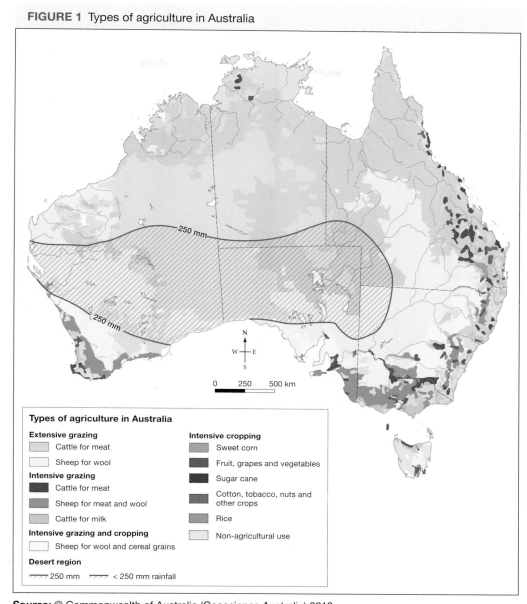

Types of agriculture in Australia

Extensive grazing
- Cattle for meat
- Sheep for wool

Intensive grazing
- Cattle for meat
- Sheep for meat and wool
- Cattle for milk

Intensive grazing and cropping
- Sheep for wool and cereal grains

Desert region
- 250 mm   < 250 mm rainfall

Intensive cropping
- Sweet corn
- Fruit, grapes and vegetables
- Sugar cane
- Cotton, tobacco, nuts and other crops
- Rice
- Non-agricultural use

Source: © Commonwealth of Australia (Geoscience Australia) 2013

FIGURE 2 Changes in agricultural land use

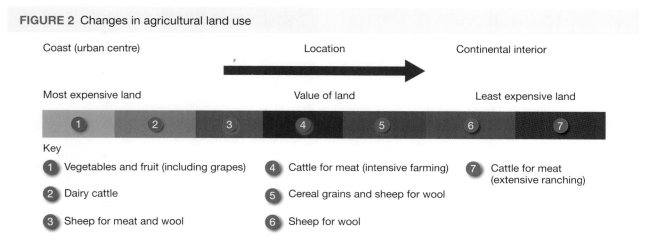

Coast (urban centre)   Location   Continental interior

Most expensive land   Value of land   Least expensive land

1 2 3 4 5 6 7

Key

1 Vegetables and fruit (including grapes)

2 Dairy cattle

3 Sheep for meat and wool

4 Cattle for meat (intensive farming)

5 Cereal grains and sheep for wool

6 Sheep for wool

7 Cattle for meat (extensive ranching)

FIGURE 3 Cattle mustering

Wheat farms

About 30 000 farms in Australia grow wheat as a major crop, and the average farm size is 910 hectares, or just over 9 square kilometres. Wheat production in Australia for 2017 was 31 million tonnes. As in other areas of the world, extensive wheat farming is found in mid-latitude temperate climates that have warm summers and cool winters, and annual rainfall of approximately 500 millimetres. In Australia, these conditions occur away from the coast in the semi-arid zone. The biome associated with this form of food production is generally open grassland, **mallee** or savanna that has been cleared for the planting of crops.

Soils can be improved by the application of fertilisers, and crop yields increased by the use of disease-resistant, fast-growing seed varieties. Wheat farms are highly mechanised, using large machinery for ploughing, planting and harvesting. The farm produce, which can amount to 2 tonnes per hectare, is sold to large corporations in local and international markets.

FIGURE 4 Harvesting wheat with a combine harvester

Mixed farms

Mixed farms combine both grazing and cropping practices. They are located closer to markets in the wetter areas, and are generally small in scale, but operate in much the same way as cattle and sheep farms.

Intensive farming

Intensive farms are close to urban centres, producing dairy, horticulture and market gardening crops. They produce milk, fruit, vegetables and flowers, all of which are perishable, sometimes bulky, and expensive to transport. The market gardens are capital- and labour-intensive, because the cost of land near the city is high, and many workers are required for harvesting.

FIGURE 5 Strawberries are typically grown in market gardens.

Plantation farming

This form of agriculture can often be found in warm, well-watered tropical places. Plantations produce a wide range of produce such as coffee, sugar cane, cocoa, bananas, rubber, tobacco and palm oil. Farms can be 50 hectares or more in size. Although many such farms in Australia are family owned, in other parts of the world they are often operated by large multinational companies. Biomes that contain plantations are mainly tropical forests or savanna, and require large-scale clearing to allow for farming. Cash returns are high, and markets are both local and global.

FIGURE 6 A banana plantation near Carnarvon, Western Australia

3.5.3 CASE STUDY: Farming around Griffith, New South Wales

Modern-day food production relies heavily on technology to create ideal farming conditions. This may involve reshaping the land to allow for large agricultural machinery and for the even distribution and drainage of water. Uneven or unreliable rainfall can be supplemented by irrigation. As a result of such changes, large areas can become important farmland.

Griffith, located in the Western Riverina region of New South Wales, is an important agricultural and food-processing centre for the region, generating more than $2 billion dollars' worth of food in 2016–17. The Riverina region has a diverse agricultural sector. The most important commodities in the region in 2016–17 were wheat ($519 million), followed by cattle ($241 million) and canola ($218 million). Oranges and rice, which are irrigated crops, were valued at $150 million and $110 million respectively.

FIGURE 7 Orange trees growing in an orchard near Griffith, New South Wales

The first European explorer to the area was John Oxley, who described the region as 'uninhabitable and useless to civilised man'. This was largely due to the lack of a suitable water supply. The construction of irrigation canals in 1906 established a reliable source of water that could be used in food production. The region has become an important food centre owing to the large-scale use of irrigation combined with suitable flat land, fertile soils and a mild climate.

To investigate the area in more detail, study the topographic map shown in **FIGURE 8**.

on Resources

Digital doc   Topographic map of Griffith (doc-11566)

Google Earth Griffith, New South Wales

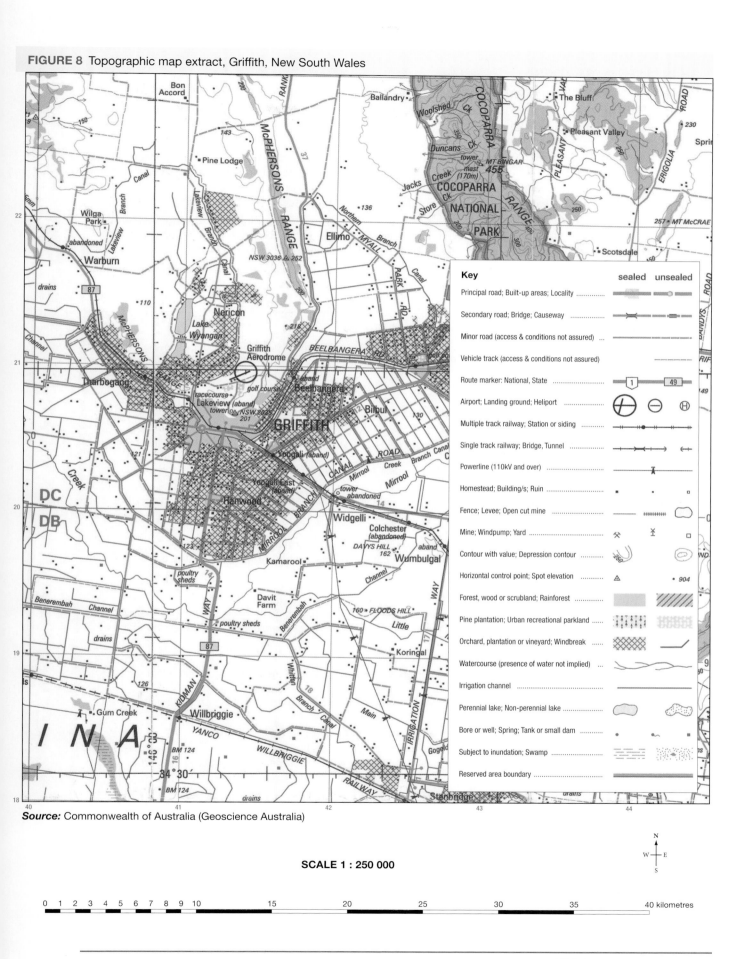

FIGURE 8 Topographic map extract, Griffith, New South Wales

Source: Commonwealth of Australia (Geoscience Australia)

SCALE 1 : 250 000

3.5 INQUIRY ACTIVITIES

1. Investigate which foods are grown closest to you. Create a map infographic showing locations and types of food grown.

   Examining, analysing, interpreting
   Classifying, organising, constructing

2. Collect information on the percentage of land used for the different forms of farming in Australia and present this data in a graph. Comment on the details shown in your graph.

   Examining, analysing, interpreting
   Classifying, organising, constructing

3. One plantation industry is palm oil production. It often has great impacts on tropical biomes; loss of habitat is one such impact. On a world map, locate major palm oil production areas and explain the implications of loss of habitat in those areas.

   Examining, analysing, interpreting

4. Various plantations in Queensland (such as pineapple, sugar cane and banana plantations) are associated with fertiliser run-off, which is affecting the Great Barrier Reef. Investigate this issue and find out what effects fertiliser has on this aquatic *environment*.

   Examining, analysing, interpreting

3.5 EXERCISES

Geographical skills key: GS1 Remembering and understanding **GS2** Describing and explaining **GS3** Comparing and contrasting **GS4** Classifying, organising, constructing **GS5** Examining, analysing, interpreting **GS6** Evaluating, predicting, proposing

3.5 Exercise 1: Check your understanding

1. **GS1** Which type of agricultural land use is closest to urban centres, and which is the furthest away?
2. **GS1** How does the *environment* in the centre of Australia affect farming types?
3. **GS1** What is the *interconnection* between climate and farm type in Australia? (*Hint:* Refer to a climate map in your atlas for ideas.)
4. **GS2** Explain why extensive, large-*scale* cattle and sheep farms are typically located in remote and arid regions of Australia.
5. **GS2** Using the **FIGURE 1** map of agriculture types in Australia, describe and explain the location of:
   (a) wheat farms
   (b) dairy farms.
6. **GS4** Are orchards and vineyards an example of intensive or extensive farming? Explain.
7. **GS3** Compare the pattern of irrigation channels and buildings in AR3919 and AR4220 in **FIGURE 8**. Suggest a reason for the differences you can see.

3.5 Exercise 2: Apply your understanding

1. **GS6** What would be the impact of flood or drought on any of the commercial methods of food production?
2. **GS6** Predict the impact of the growth of Australian capital cities on the *sustainability* of surrounding market gardens.
3. **GS2** Why is much of Australia's food production available for export?
4. **GS6** It used to be said that Australia's economy 'rode on the sheep's back'. What do you think this means, and do you think it is still true today?
5. **GS6** What types of *environment* might have existed in the Griffith area when Oxley first arrived?
6. **GS5** Study **FIGURE 8**.
   (a) Identify and name a possible source of irrigation water on the map.
   (b) How is water moved around this area? (*Hint:* Follow the blue lines.)
   (c) Using the contour lines and spot heights as a guide, estimate the average elevation of the map area.
   (d) What is the importance of topography (the shape of the land) to irrigation?
   (e) What types of farming are found at the following *places?*
      i. GR410195
      ii. GR413220
   (f) Approximately what percentage of the map area is irrigated?
   (g) Within Griffith there are many factories that process raw materials, such as rice mills, wineries and juice factories. What would be the advantages and disadvantages of locating processing factories close to growing areas?

Try these questions in learnON for instant, corrective feedback. Go to www.jacplus.com.au.

3.6 SkillBuilder: Describing patterns and correlations on a topographic map

What are patterns and correlations on a topographic map?

A pattern is a sequence in which features are distributed or spread. A correlation shows how two or more features are interconnected — that is, the relationship between the features. Patterns and correlations in a topographic map can show us cause-and-effect connections.

Select your learnON format to access:

- an overview of the skill and its application in Geography (Tell me)
- a video and a step-by-step process to explain the skill (Show me)
- an activity and interactivity for you to practise the skill (Let me do it)
- questions to consolidate your understanding of the skill.

Resources

Video eLesson Describing patterns and correlations on a topographic map (eles-1729)

Interactivity Describing patterns and correlations on a topographic map (int-3347)

3.7 Indigenous Australians' food security over time

3.7.1 Caring stewards of the land

The sustainable land and resource management practices of Aboriginal and Torres Strait Islander peoples carried out over many thousands of years ensured food security for the people and respect for the lands, waterways, lakes and marine environments that sustained them. At the time of European occupation in 1788, most Indigenous Australians were hunters and gatherers. However, some nations had abundant food supplies in their regions and were able to largely settle in one place. In all cases their deep knowledge and close association with the land allowed for sustainable management of the ecosystems and biomes in which they lived. The 'world view' that describes this sustainable lifestyle is called an 'earth-centred' approach. This means people's interaction with the environment is one of caring stewardship.

3.7.2 Sourcing food

Indigenous Australian peoples sourced their foods from a wide range of uncultivated plants and wild animals, with some estimates suggesting there were up to 7000 different sources of food. The composition of the food was greatly influenced by both the season and geographic location of the community region.

In Aboriginal communities there was a division of labour among men, women and children. Food sources based on cereals, fruits and vegetables were collected or gathered daily by women and children. Men were involved more in hunting for game and fishing, as well as in wider-scale land management using fire.

To ensure food security, communities developed a range of sustainable food-gathering techniques. For example, some seeds from gathered plants were left behind to allow for new growth, and a few eggs were always left in nests to hatch. This ensured that species would survive and communities could expect to find food in the same place in the future. **FIGURE 1** provides details of traditional food types from both tropical and temperate regions of Australia, including arid and desert regions.

FIGURE 1 A selection of different foods and water resources

Cereal foods: Grass seeds from the clover fern were ground to form flour for damper. Many other seed types were similarly treated.

Fruit and vegetables: Fruits, berries, orchids and pods were available, depending on the region and seasonal availability (for example, sow thistle, lilly pilly, pigface fruit, kangaroo apple, wild raspberry, quandong and native cherry) as well as wild figs, plums, grapes and gooseberries. Also eaten were plant roots such as bull rushes, yams and bulbs; the heart of the tree fern and the pith of the grass tree; and the blister gum from wattles, native truffles and mushrooms.

Eggs: Emu, duck, pelican and many other birds' eggs were eaten.

Meat: Meats included insects such as the larval stage of the cossid moth or witchetty grub and the Bogong moth, honey ants, native bees and their honey, and scale insects; animals such as kangaroos, emus, eels, crocodiles, sea turtles, snakes, goannas and other lizards; and birds such as ducks, gulls and pelicans.

Fish and shellfish: Freshwater fish, such as perch, yabbies and mussels in creeks and rock holes, and all varieties of saltwater fish were caught.

Medicines: Over 120 native plants were used as sedatives, ointments, diarrhoea remedies, and cough and cold palliatives as well as for many other known treatments.

Water: Water was obtained from rivers, lakes, rock holes, soaks, beds of intermittent creeks and dew deposited on surfaces. Moisture obtained from foods such as tree roots and leaves also provided water.

Torres Strait Islander peoples

Torres Strait Islander peoples' food sources, both historically and today, are based on fishing, horticulture and inter-island trading activities. Torres Strait Islander peoples have a profound understanding of the sea, including its tides and sea life. While their food sources vary from island to island, their lifestyle can be best described as subsistence agriculture with seafood, garden foods and other produce stored and preserved for both local use and trade.

FIGURE 2 Cooking bush food in a traditional Kup Murrie, or ground oven

3.7.3 The use of fire

The use of fire was a significant aspect of Indigenous Australian peoples' land management. What has been described as the 'park-like' landscape of the Australian bush was purposely created by clearing forest in controlled burns using fire sticks. After the fires, new plant growth with tender shoots attracted all types of birds and animals to the area. The grassland areas that resulted from the controlled burning of the landscape became ideal places to hunt kangaroos. Burning also flushed animals out into the open where they could be speared (see **FIGURE 3**). Indigenous Australian peoples' use of fire had to be carefully managed as part of their efforts to ensure food security. As such, it was a sustainable practice based on a sound knowledge of fire control and the environment.

FIGURE 3 Using fire for hunting and to manage the land

3.7.4 The Bogong moth: a past food source

While there were many other sources of food for Indigenous communities that lived near the south-eastern Australian highlands, the Bogong moth was a particularly important seasonal specialty. The Bogong moths, which lived in the ground as larvae in Queensland, migrated in millions to the south-eastern highlands to seek out cool, rocky overhangs and crevices where they could sleep through the long, hot summer months, surviving off the fat in their bodies (see **FIGURE 4**).

The Bogong moths were a rich source of fat and protein for Indigenous peoples who lived adjacent to the highlands of Victoria and New South Wales. Many culture groups would migrate from the valleys and foothills into the highlands and set up camps for the feasting ceremony. They would smoke out the moths, collecting them by the thousands to be cooked over hot rocks. In addition to savouring this important seasonal food source, making the annual pilgrimage to the high country presented these groups with an important opportunity to interact socially, participate in ceremonies and to arrange inter-community marriages.

FIGURE 4 Massed Bogong moths on a rock face

3.7.5 Eel farming by the Gunditjmara people

The home of the Gunditjmara people, the Budj Bim Cultural Landscape, is the site of one of Australia's largest ancient aquaculture systems. This area, which is part of the Mount Eccles National Park near Portland in Victoria, shows evidence of a large, permanent settlement of stone huts and channels used for farming and the local trade of eels (see **FIGURE 5**). The Gunditjmara people managed this landscape by digging channels and constructing weirs to bring water and young eels from Darlot Creek to local ponds and wetlands. Woven baskets placed at the weirs were used to harvest the mature eels. The area provided an abundance of food, ensuring food security for all.

Following European occupation of the area in the 1830s, the Gunditjmara people fought for their lands in the Eumerella Wars, which lasted for more than 20 years. By the 1860s the remaining Gunditjmara people were displaced to a government mission at Lake Condah. The mission lands were returned to the Gunditjmara people in 1987. The Deen Maar Indigenous Protected Area (IPA) was declared in 1999 and the area was listed on the Australian National Heritage register in 2004. The Budj Bim Cultural Landscape was added to the World Heritage List in July 2019.

Today the Gunditjmara people, as part of the Winda-Mara Aboriginal Corporation, manage the 248-hectare Darlots Creek (Killara), which flows from Lake Condah in the Budj Bim Cultural Landscape.

The wetlands and manna gum woodlands have been largely re-established through works to control weeds and feral animals. There are also prospects of restarting the eel aquaculture industry as a sustainable business. To further eco-tourism, boardwalks have been built, signage put in place, and a range of tours of the wetlands, lakes and woodlands are offered by Indigenous guides. These tours focus on the past way of life, and examine rebuilt channels, weirs and eel traps, and traditional sourcing of food stocks.

FIGURE 5 Remains of Aboriginal stone eel traps at Lake Condah

3.7.6 Food security issues in remote Indigenous communities

Historically, Aboriginal and Torres Strait Islander peoples managed their food and water resources sustainably to prevent food insecurity. With European occupation, however, over time they were largely displaced from their lands and traditional methods of sourcing food were altered. The loss of traditional hunting, gathering and fishing areas played a role in limiting people's access to nutritious and fresh sources of food.

In recent times, with the passing of the *Native Title Act 1993* and involvement of federal and state governments, members of Indigenous communities have been able to re-establish connections with their lands through various collaborative land- and water-management projects. However, many Indigenous peoples living in remote areas still experience food insecurity due to changes to traditional food sourcing methods and a range of social issues stemming from displacement and historical government policies. Related health issues include a higher risk of respiratory diseases, cardiovascular disease, diabetes, chronic kidney disease and mental health issues than that faced by non-Indigenous Australians. The importance of access to fresh, healthy food supplies cannot be understated as a factor in redressing this imbalance.

Why is food security difficult in remote communities?

Over 80 per cent of Aboriginal and Torres Strait Islander peoples reside in cities and non-remote areas across Australia, and nearly 20 per cent live in remote and very remote areas (see **FIGURE 6**).

FIGURE 6 Distribution of remote Aboriginal and Torres Strait Islander communities

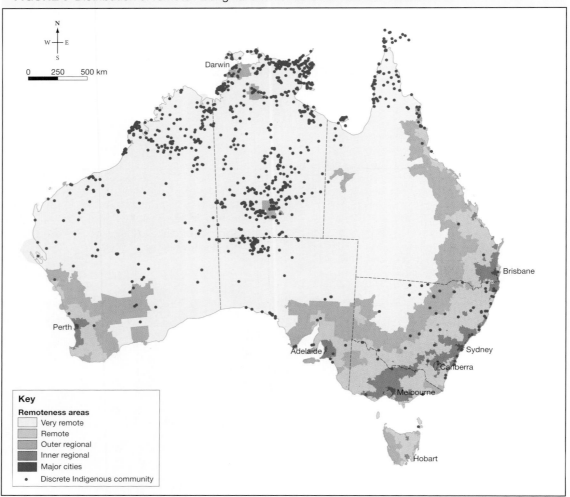

Source: Australian Bureau of Statistics

Access to affordable and nutritious food is an ongoing problem in remote communities. These remote areas have a high cost of living and suffer from a general unavailability of fresh fruit and vegetables as well as other high-quality foods, due to the associated costs of transportation and storage. The Australian government's 2016 National Workforce Action Plan Report found that 31 per cent of Aboriginal and Torres Strait Islander peoples living in remote areas were likely to be in households that had run out of food or could not afford to buy food, compared to 20 per cent for non-Indigenous people. As the main source of food is often limited to the local general store, there are ongoing risks of food insecurity and consequent health issues related to limited choices of diet.

Measures to alleviate food insecurity

The Australian government's 2013 National Food Plan sought to address matters such as food security, the affordability and quality of food, and the sustainability of Australian food production. Measures that have been funded by federal and state governments that address food insecurity for remote Indigenous communities include:

- improving the food supply chain to remote areas
- establishing and improving community stores in remote areas through initiatives such as the Outback Stores program
- education campaigns about nutrition guidelines and healthy food choices
- grants to support the establishment or improvement of community food initiatives, such as farmers' markets, food cooperatives, food hubs, community gardens and city farms
- food subsidies for fruit and vegetable programs to improve health and nutrition for Indigenous children
- funding for the Stephanie Alexander Kitchen Garden National Program to develop gardens in schools
- improving skills of Indigenous health workers to enable them to have greater influence on food security through addressing health promotion and nutrition at the local level.

FIGURE 7 Papunya Community Store, located 240 kilometres north-west of Alice Springs

Funding has also been made available to various industries that relate to Indigenous peoples' connection to the land. Industry ventures have included park and ranch management to further eco-tourism, and aquaculture. These ventures can also provide food and are a source of revenue for further community development, employment and training opportunities. They can be seen as developing a sense of pride as well as prospects for future generations. They are in keeping with Indigenous peoples' roles as caretakers and traditional custodians of the land.

One such initiative is Fish River Station in the Daly River region of the Northern Territory, home to an array of wildlife, plant life and important traditional food sources for Indigenous peoples. The partnership between government, conservation organisations and the Indigenous Land Corporation aims to preserve this unique environment for future generations, whilst providing employment for Indigenous people and opportunities to reconnect with this culturally significant land.

DISCUSS
What can we learn from the land and resource management practices of Indigenous Australians in relation to food security?

[Intercultural Capability]

FIGURE 8 Fish River Station rangers Desmond Daly and Jeff Long patrol 178 000 hectares of land.

 Resources

🔗 **Weblinks** Indigenous food sources

Fish River Station

Outback Stores

3.7 INQUIRY ACTIVITY

Use the weblinks provided in the Resources tab to explore and learn more about **Indigenous food sources**, the **Outback Stores** project and **Fish River Station**. With a partner, choose one of these topics and create an A4 infographic poster summarising your findings.
Examining, analysing, interpreting
Classifying, organising, constructing

3.7 EXERCISES

Geographical skills key: GS1 Remembering and understanding **GS2** Describing and explaining **GS3** Comparing and contrasting **GS4** Classifying, organising, constructing **GS5** Examining, analysing, interpreting **GS6** Evaluating, predicting, proposing

3.7 Exercise 1: Check your understanding

1. **GS1** What was the division of labour for men, women and children when sourcing food?
2. **GS1** How did Indigenous Australian peoples use fire to source food?
3. **GS1** Who are the traditional owners of the Budj Bim region of Victoria?
4. **GS2** Why did Aboriginal and Torres Strait Islander peoples lose access to many traditional food sources after European occupation?
5. **GS2** How did the Indigenous Australian peoples access the Bogong moth as a food source?
6. **GS1** What are some of the increased health risks experienced by Indigenous Australians as compared with non-Indigenous Australians?

3.8 Rice — an important food crop

3.8.1 The importance of rice

Rice is the seed of a semi-aquatic grass. In warm climates, in more than 100 countries, it is cultivated extensively for its edible grain. Rice is one of the most important staple foods of more than half of the world's population, and it influences the livelihoods and economies of several billion people. In Asia, rice provides about 49 per cent of the calories and 39 per cent of the protein in people's diet. In 2017, approximately 760 million tonnes of rice were produced worldwide.

FIGURE 1 shows that the largest concentration of rice is grown in Asia. About 132 million hectares are cultivated with this crop, producing 88 per cent of the world's rice. Of this, 48 million hectares and 31 per cent of the global rice crop are in South-East Asia alone.

Countries with the largest areas under rice cultivation are India, China, Indonesia, Bangladesh, Thailand, Vietnam, Myanmar (Burma) and the Philippines, with 80 per cent of the total rice area.

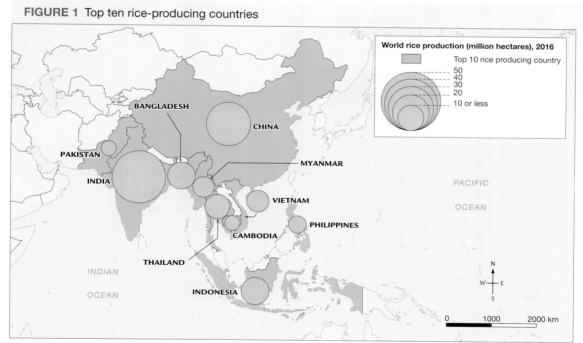

FIGURE 1 Top ten rice-producing countries

Source: WorldAtlas.com, http://www.worldatlas.com/articles/the-countries-producing-the-most-rice-in-the-world.html
(*Note:* most current data available)

3.8.2 Factors affecting rice production in Asia

Climate and topography

Rice can be grown in a range of environments that are hot or cool, wet or dry. It can be grown at sea level on coastal plains and at high altitudes in the Himalayas. However, ideal conditions in South-East Asia include high temperatures, large amounts of water, flat land and fertile soil.

In Yunnan Province, China, the mountain slopes have been cultivated in terraced rice paddies by the Hani people for at least 1300 years (see **FIGURE 2**). The terraces stop erosion and surface run-off.

FIGURE 2 Rice terraces in Yunnan Province, China. These terraces are at an elevation of 1570 metres.

Irrigation

Traditional rice cultivation involves flooding the paddy fields (*padi* meaning 'rice plant' in Malay) for part of the year. These fields are small, and earth embankments (*bunds*) surround them. Rice farmers usually plant the seeds first in little seedbeds and transfer them into flooded paddy fields, which are already ploughed. Canals carry water to and from the fields. Houses and settlements are often located on embankments or raised islands near the rice fields.

Approximately 45 per cent of the rice area in South-East Asia is irrigated, with the largest areas being in Indonesia, Vietnam, the Philippines and Thailand. High-yielding areas of irrigated rice can also be found in China, Japan and the Republic of Korea. Because water is available for most of the year in these places, farmers can grow rice all year long. This intensive scale of farming can produce two and sometimes three crops a year.

Upland rice is grown where there is not enough moisture to nurture the crops; an example of such cultivation takes place in Laos. This method produces fewer rice varieties, since only a small amount of nutrients is available compared to rice grown in paddy fields.

FIGURE 3 Planting rice in paddy fields in north-east Thailand

padi (rice plant)

bund (embankment)

Impacts on potential yield

Rice yields can be limited if any of the following conditions exist:
- poor production management
- losses caused by weeds (biotic factor)
- pests and diseases (biotic factor)
- inadequate land formation and irrigation water
- inadequate drainage that leads to a build-up of salinity and alkalinity.

Technology

Agricultural biotechnology, especially in China, has produced rice that is resistant to pests. There are also genes for herbicide resistance, disease resistance, salt and drought tolerance, grain quality and photosynthetic efficiency. Genetic engineering may be the way of the future in rice cultivation in some parts of the world.

In the Philippines, through cross-breeding rather than genetic engineering, a new strain of rice has been developed that grows well in soils lacking phosphorus (see **FIGURE 4**). This could also have a significant positive impact on crop yields.

Environmental issues

Increasing temperatures, caused by global warming, may be causing a drop in rice production in Asia, where more than 90 per cent of the world's rice is produced and consumed. The Food and Agriculture Organization of the United Nations (FAO) has found that in six of Asia's most important rice-producing countries — China, India, Indonesia, the Philippines, Thailand and Vietnam — rising temperatures over the past 25 years have led to a 10–20 per cent decline in rice output.

Scientists state that if rice production methods cannot be changed, or if new rice strains able to withstand higher temperatures cannot be developed, there will be a loss in rice production over the next few decades as days and nights get hotter. People may need to turn to a new staple crop.

Rice growing is eco-friendly and has a positive impact on the environment. Rice fields create a wetland habitat for many species of birds, mammals and reptiles. Without rice farming, wetland environments created by flooded rice fields would be vastly reduced.

FIGURE 4 Rice demonstration plots at the International Rice Research Institute in the Philippines.

3.8.3 Factors affecting rice production in Australia

Climate and topography

Eighty per cent of rice produced in Australia consists of temperate varieties that suit climates with high summer temperatures and low humidity. Large-scale production of rice is carried out in the Murrumbidgee and Murray valleys of New South Wales; the production process is sophisticated.

Sowing and irrigation

In Australia, rice grows as an irrigated summer crop from September to March. Most of it is sown by aircraft rather than planted by hand. Experienced agricultural pilots use satellite guidance technology to broadcast seed accurately over the fields.

Before sowing, the seed is soaked for 24 hours and drained for 24 hours, leaving a tiny shoot visible on the seed. Once sown, it slowly settles in the soft mud, and within three to four days each plant develops a substantial root system and leaf shoot. After planting, fresh water is released from irrigation supply channels to flow across each paddy field until the rice plants are well established.

Most countries grow rice as a **monoculture**, whereas Australian rice grows as part of a unique farming system. Farmers use a **crop rotation** cycle across the whole farm over four to five years. This means that the growers have other agricultural enterprises on the farm as well as rice. This system, designed for efficiency, sustainability and safety, means Australian growers maintain water savings, and have increased soil nutrients, higher yields and much healthier crops.

Once Australian rice growers harvest their rice, they use the subsoil moisture remaining in the soil to plant another crop — either a wheat crop or pasture for animals. This form of rotation is the most efficient in natural resource use and agricultural terms.

Pests and diseases

Rice bays (areas contained by embankments — see **FIGURE 5**) are treated with a chemical application, which prevents damage by pests and weeds. Without this treatment, crop losses would be extensive. In the last 100 days before harvesting, the rice plant has no chemical applications, so that when it is harvested, it is virtually chemical free.

FIGURE 5 Murrumbidgee irrigation area rice bays

Technology

Most farms use laser-guided land-levelling techniques to prepare the ground for production. This gives farmers precise control over the flow of water on and off the land. Such measurement strategies have contributed to a 60 per cent improvement in water efficiency. Most of the equipment used on rice farms is fitted with computer-aided devices, such as GPS (global positioning systems), GIS (geographical information systems) and remote sensing. Australian rice growers are the most efficient and productive in the world.

FIGURE 6 Harvesting rice near Griffith, New South Wales

Environmental issues

The rice industry encourages biodiversity enhancement and greenhouse gas reduction strategies. Some farms in southern New South Wales are avoiding the use of chemical fertilisers and pesticides by converting farms to biodynamic practices; for example, they have avoided salinity by planting red gums.

Drought conditions from 2002–09 and 2013–16 in the Murray-Darling Basin led to reduced water allocations for crop irrigation in the Riverina area. The subsequent 2015–16 drought years, in particular, led to a reduction in rice production. In the context of the health of our waterways, there is ongoing debate about the impact of irrigation-dependent farming and how best to sustainably balance environmental concerns and our food production needs now and for the future.

 Explore more with my**World**Atlas

Deepen your understanding of this topic with related case studies and questions.
- Investigating Australian Curriculum topics > Year 9: Biomes and food security > **Rice**

On Resources

 Interactivity   How is rice grown? (int-3322)
 Weblink   Terraced rice

3.8 INQUIRY ACTIVITIES

1. Investigate two different rice-growing *places* in Asia and describe the reasons for the different *environments*.   **Examining, analysing, interpreting**
2. Investigate an example of an Australian rice farm and outline its yearly rice-growing cycle.
                                                                       Examining, analysing, interpreting
3. Research the *interconnection* between rice growing and the Murray River for ensuring a *sustainable environment*.   **Examining, analysing, interpreting**

3.8 EXERCISES

Geographical skills key: GS1 Remembering and understanding **GS2** Describing and explaining **GS3** Comparing and contrasting **GS4** Classifying, organising, constructing **GS5** Examining, analysing, interpreting **GS6** Evaluating, predicting, proposing

3.8 Exercise 1: Check your understanding

1. **GS5** Refer to **FIGURE 1**. Which two countries produce most of the world's rice?
2. **GS2** Name the other top-ten rice-producing countries in the world. What is the geographical location of these *places*?
3. **GS1** What is meant by the term *monoculture*?
4. **GS1** What is crop rotation?
5. **GS2** Explain why *places* in Asia are ideally suited to rice growing.

3.8 Exercise 2: Apply your understanding

1. **GS2** Explain the benefits of a crop-rotation system.
2. **GS2** Explain how mountain slopes in China have been *changed* to accommodate rice production.
3. **GS6** Explain the *environmental* issues that may affect future rice production.
4. **GS3** Describe and explain the similarities and differences between the rice cultivation methods used in Asia and Australia.
5. **GS6** Predict how technology will influence *changes* to rice cultivation in both Asia and Australia.

Try these questions in learnON for instant, corrective feedback. Go to www.jacplus.com.au.

3.9 Cacao — a vital cash crop

3.9.1 Global cacao production feeds global chocolate demand

Chocolate is made from the beans of the cacao plant. The chocolate that is produced from these beans might come from cacao grown in Ghana, Mexico, Malaysia or Indonesia. Look on the wrapper next time you eat a chocolate bar!

Chocolate has been eaten or drunk by people for 4000 years. A recent study showed that 91 per cent of females and 87 per cent of males consume chocolate products in places such as Great Britain, Australia, Switzerland and the United States. However, rising disposable incomes and changing tastes will continue to change the scale of production, both overseas and locally; people in places such as India and China are now eating more chocolate.

The cacao tree is a native of the Amazon Basin and other tropical areas of South and Central America, where wild varieties still grow in the forests. Many countries now grow cacao, but the main places are:

- West Africa — Ghana, Nigeria and Cote D'Ivoire
- South America — Brazil and Ecuador
- Asia — Malaysia and Indonesia.

Malaysia and Indonesia, where cacao is a relatively new crop, are becoming increasingly important growing areas.

FIGURE 1 The main cacao-growing regions of the world

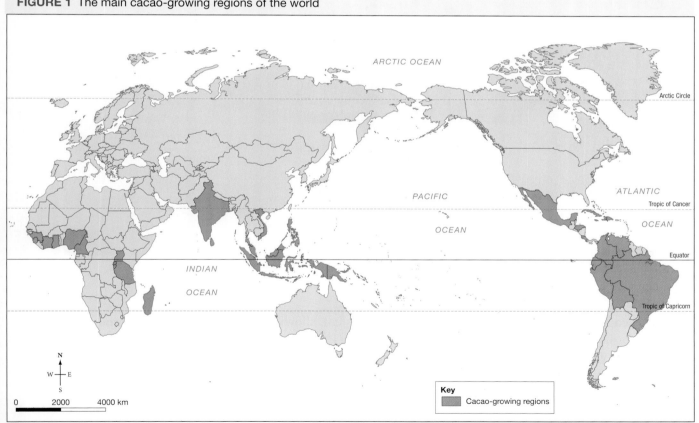

Source: Spatial Vision

3.9.2 Influences on the growth and production of cacao

Environmental factors

Most of the world's cacao is grown in a narrow belt between 10 degrees north and 10 degrees south of the equator. Cacao trees grow well in humid tropical climates with regular rainfall and a short dry season. The trees need temperatures between 21 °C and 23 °C and rainfall of 1000 to 2500 millimetres per year. The soil must drain well but have good moisture-holding capacity. The trees cannot tolerate tropical sun and must be grown in the shade of other trees, such as palms and rubber plants. Although cacao plants can reach a height of 12 metres, most are only six to seven metres tall. Growth is very fast, and the plant can flower and fruit two to three years after planting.

FIGURE 2 A cacao bean and seeds

Economic factors

Cacao is a **cash crop**, grown mostly in an **agroforestry** system, allowing for biodiversity and **income diversity** for families.

Around the world, six million cacao farmers — and 40 to 50 million people in total — depend on cacao for their livelihood. For the past century, demand has grown by three per cent per year.

Cacao beans are an important export for West African nations such as Ghana and the Cote d'Ivoire. These countries are the source of more than 70 per cent of the world's cocoa. Cacao beans are traded on the world market and their price can change daily, depending on supply and demand around the world. For example, too many beans on the world market can cause prices to drop, leaving farmers without the cash they need to cultivate their crops, and this ultimately lowers the supply. Adverse weather or tree disease can shrink supply as well.

FIGURE 3 A cacao farmer from Ghana carrying cacao pods

Labour

Cacao is one of the world's most labour-intensive crops. Much of the work is done by hand on a daily basis. The flowers are often pollinated by hand and defective pods are removed to allow the plant to put more energy into good ones.

Cultural factors

For the farmers themselves, the work of cacao production can be a collective effort and very much a part of family and village life. Production stages such as 'turning the beans' — an involved process that takes place over several days — present an opportunity to commune with other farming families. But beyond this, cacao and chocolate are an important part of the cultural practices of many communities throughout the world.

In Oaxaca, Mexico, traditional healers called *curanderos* give chocolate drinks to cure bronchitis. They also plant cacao beans in the earth to ward off evil forces and heal those who have *espanto* — sickness caused by fright. Children drink chocolate for breakfast to protect them against stings from scorpions or bees. In Australia and many other countries, chocolate has become an integral part of cultural events such as Easter and Christmas. Indeed, in Western society, chocolate is very much a part of everyday life, which contributes to the ever-increasing global demand for this special crop.

What is the future for chocolate?

Consumer demand for chocolate is on the rise, but the cacao tree is under threat from pests, fungal infections, climate change, and farmers' lack of access to fertilisers and other products that enhance yields. In West Africa, there are efforts to train farmers in organic, sustainable farming practices.

Global consumption is increasing, especially of darker, more cocoa-heavy varieties. Research is underway to develop hardier trees that can produce bigger yields while still making tasty chocolate. Fairtrade arrangements are improving the lives of farmers, increasing their income and helping them replace old trees and equipment (see subtopic 9.7). Through such arrangements, it is hoped that cacao production can be sustainably managed, providing ongoing income and contributing to the wellbeing of the many millions of people within the cacao-producing communities throughout the world.

3.9 INQUIRY ACTIVITY

Research Fairtrade International and learn how this organisation has enabled the *sustainability* of cacao in many *places* in the world. Create an infographic outlining your key findings.

Examining, analysing, interpreting
Classifying, organising, constructing

3.9 EXERCISES

Geographical skills key: GS1 Remembering and understanding **GS2** Describing and explaining **GS3** Comparing and contrasting **GS4** Classifying, organising, constructing **GS5** Examining, analysing, interpreting **GS6** Evaluating, predicting, proposing

3.9 Exercise 1: Check your understanding

1. **GS1** List the main cacao-producing *places* in the world.
2. **GS2** Refer to **FIGURE 1** to describe the geographical location of cacao-producing *places*. Refer to lines of latitude in your response.
3. **GS1** Outline the main *environmental* factors necessary for successful cacao-tree growth.
4. **GS1** What *environmental* factors suit cacao growing in the Daintree region of North Queensland?
5. **GS3** Explain the difference between a cash crop and a subsistence crop.

3.9 Exercise 2: Apply your understanding

1. **GS2** Explain the significance of chocolate in different cultures throughout the world.
2. **GS2** Explain how world cacao-bean prices can affect a cacao farmer's income.
3. **GS2** Why is cacao considered a labour-intensive crop?
4. **GS2** Describe the global consumption of chocolate and how it is *changing*.
5. **GS6** It has been suggested that there could be a crisis in chocolate production in future years, with chocolate becoming rare and very expensive. Should money be spent on research to produce hardier cacao trees with bigger yields just to satisfy the chocolate desires of the Western world? Should money be spent on other types of agriculture? Outline your views on this issue, considering all you have learned in this subtopic.

Try these questions in learnON for instant, corrective feedback. Go to www.jacplus.com.au.

3.10 SkillBuilder: Constructing ternary graphs

What are ternary graphs?

Ternary graphs are triangular graphs that show the relationship or interconnection between features. They are particularly useful when a feature has three components and the three components add up to 100 per cent.

Select your learnON format to access:

- an overview of the skill and its application in Geography (Tell me)
- a video and a step-by-step process to explain the skill (Show me)
- an activity and interactivity for you to practise the skill (Let me do it)
- questions to consolidate your understanding of the skill.

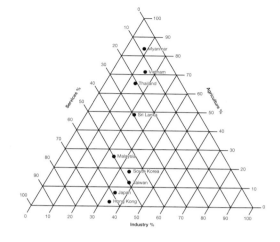

Resources

🎬 **Video eLesson** Constructing ternary graphs (eles-1728)

🧩 **Interactivity** Constructing ternary graphs (int-3346)

3.11 Thinking Big research project: Subsistence living gap-year diary

SCENARIO

As an end-of-school gap-year experience, you have been given the opportunity to live with an indigenous community that relies on traditional subsistence approaches to food security such as nomadic herding, hunting-gathering or shifting agriculture. Your gap-year diary will help you remember forever the people you've met and their distinctive ways of managing their environment.

Select your learnON format to access:

- the full project scenario
- resources to guide your project work
- an assessment rubric.

Resources

 ProjectsPLUS Thinking Big research project: Subsistence living gap-year diary (pro-0189)

3.12 Review

3.12.1 Key knowledge summary
Use this dot point summary to review the content covered in this topic.

3.12.2 Reflection
Reflect on your learning using the activities and resources provided.

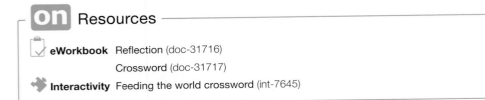

on Resources

☑ **eWorkbook**   Reflection (doc-31716)

            Crossword (doc-31717)

🧩 **Interactivity**   Feeding the world crossword (int-7645)

KEY TERMS

agribusiness business set up to support, process and distribute agricultural products

agroforestry the use of trees and shrubs on farms for profit or conservation; the management of trees for forest products

arable describes land that can be used for growing crops

biofuel fuel that comes from renewable sources

cash crop a crop grown to be sold so that a profit can be made, as opposed to a subsistence crop, which is for the farmer's own consumption

crop rotation a procedure that involves the rotation of crops, so that no bed or plot sees the same crop in successive seasons

extensive farm farm that extends over a large area and requires only small inputs of labour, capital, fertiliser and pesticides

Green Revolution a significant increase in agricultural productivity resulting from the introduction of high-yield varieties of grains, the use of pesticides and improved management practices

horticulture the practice of growing fruit and vegetables

hybrid plant or animal bred from two or more different species, sub-species, breeds or varieties, usually to attain the best features of the different stocks

income diversity income that comes from many sources

innovation new and original improvement to something, such as a piece of technology or a variety of plant or seed

intensive farm farm that requires a lot of inputs, such as labour, capital, fertiliser and pesticides

mallee vegetation areas characterised by small, multi-trunked eucalypts found in the semi-arid areas of southern Australia

monoculture the cultivation of a single crop on a farm or in a region or country

per capita per person

sustainable describes the use by people of the Earth's environmental resources at a rate such that the capacity for renewal is ensured

undulating describes an area with gentle hills

---

4 The impacts of global food production

4.1 Overview

Our planet works hard to feed the ever-growing human population ... but at what cost?

4.1.1 Introduction

The world's biomes support human life. We depend on them for food, water, fibres, fuel and wood. We also rely on the services they provide, such as air purification, climate regulation and spiritual and aesthetic comfort. However, our ecosystems are under threat; increasing population and demand for food have altered many of our biomes and we have seen a significant decline in species biodiversity. There is a growing concern that, unless such problems are addressed, we may not be able to enjoy food security in the future.

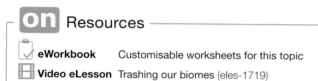 Resources

☑ **eWorkbook**   Customisable worksheets for this topic

▦ **Video eLesson** Trashing our biomes (eles-1719)

LEARNING SEQUENCE

4.1 Overview
4.2 Food production's effect on biomes
4.3 Changing our forest biome
4.4 **SkillBuilder:** GIS — deconstructing a map   online only
4.5 Paper profits, global losses?
4.6 Depleting our bountiful ocean biome
4.7 **SkillBuilder:** Interpreting a geographical cartoon   online only
4.8 Losing the land
4.9 The effects of farmland irrigation
4.10 Diminishing global biodiversity
4.11 Does farming cause global warming?
4.12 **Thinking Big research project:** Fished out! PowerPoint   online only
4.13 **Review**   online only

To access a pre-test and starter questions and receive immediate, **corrective feedback** and **sample responses** to every question, select your learnON format at www.jacplus.com.au.

4.2 Food production's effect on biomes

4.2.1 What is our biophysical world?

Food is essential to human life, and over the past centuries we have been able to produce more and more food to feed our growing population. While technology has enabled us to increase production, it has come at a price. Large-scale clearing of our forests, the overfishing of our oceans, and the constant overuse of soils has resulted in a significant deterioration of our biophysical world.

Planet Earth is made up of four spheres: the atmosphere, lithosphere, hydrosphere and biosphere (see **FIGURE 1**).

FIGURE 1 The Earth's four spheres

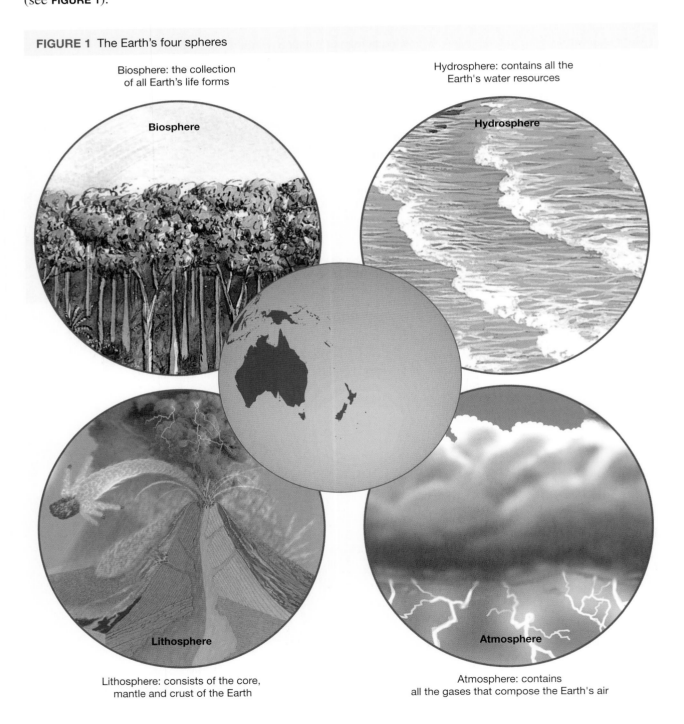

Biosphere: the collection of all Earth's life forms

Biosphere

Hydrosphere: contains all the Earth's water resources

Hydrosphere

Lithosphere

Lithosphere: consists of the core, mantle and crust of the Earth

Atmosphere

Atmosphere: contains all the gases that compose the Earth's air

All these spheres are interconnected and make up our natural or **biophysical environment**. For example, rain falling from a cloud (atmosphere) may soak into the soil (lithosphere) or flow into a river (hydrosphere) before being taken up by a plant or animal (biosphere) where it may evaporate and return to the atmosphere. (It is interesting to note that 97 per cent of the Earth's water is found in salty oceans, and the remainder as vapour in the atmosphere and as liquid in **groundwater**, lakes, rivers, glaciers and snowfields.)

Natural events, such as storms or earthquakes, or human activities can create changes to one or all of these spheres. The production of food, whether from the land or sea, has the potential to change the natural environment and, in doing so, increases the likelihood of food insecurity. **TABLE 1** shows how food production can affect the biophysical world. As can be seen, activities such as land clearing and **irrigation** can have impacts on all four of the Earth's spheres.

TABLE 1 How food production affects the biophysical world

Activities	Atmosphere	Lithosphere	Biosphere	Hydrosphere
Clearing of native vegetation for agriculture	x	x	x	x
Overgrazing animals		x	x	x
Overusing irrigation water, causing saline soils		x	x	x
Burning forests to clear land for cultivation	x	x	x	x
Run-off of pesticides and fertilisers into streams		x	x	x
Producing greenhouse gases by grazing animals and rice farming	x			
Changing from native vegetation to cropping		x	x	x
Withdrawing water from rivers and lakes for irrigation	x	x	x	x
Overcropping soils		x	x	x
Overfishing some species			x	

4.2.2 What has happened to our biophysical world?

Currently, the world produces enough food to feed all 7.7 billion people. We produce 17 per cent more food per person than was produced 30 years ago, and the rate of food production has been greater than the rate of population growth. This has been the result of: improved farming methods; the increased use of fertilisers and pesticides; large-scale irrigation; and the development of new technologies, ranging from farm machinery to better quality seeds.

There have been many benefits associated with this change, especially in terms of human wellbeing and economic development. However, at the same time, humans have changed the Earth's biomes more rapidly and more extensively than in any other time period. The loss of biodiversity and degradation of land and water (which are essential to agriculture) is not sustainable. With an expected population of 9.7 billion in 2050, it has been estimated that food production will need to increase by approximately 70 per cent. The global distribution of environmental risks associated with food production can be seen in **FIGURE 2**.

FIGURE 2 State of the world's land and water resources for food and agriculture

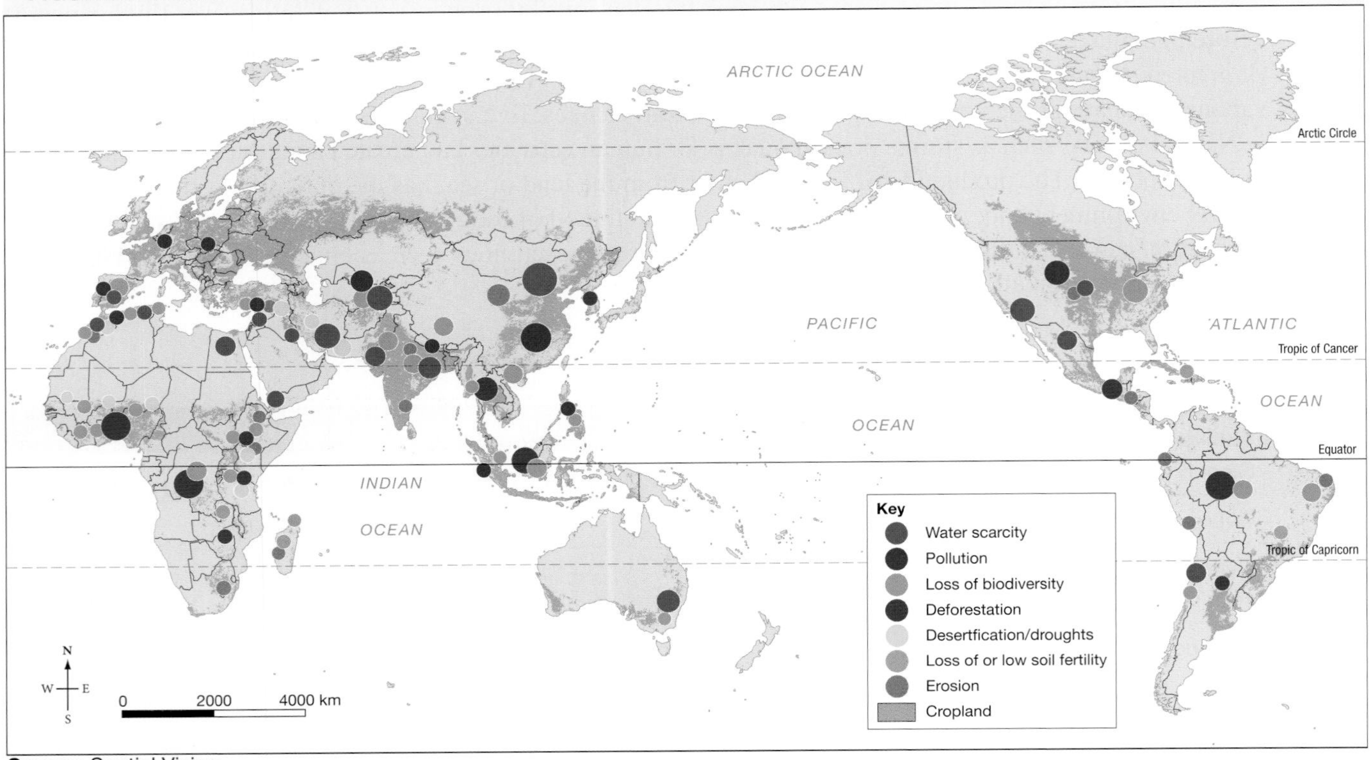

Source: Spatial Vision

Resources

Interactivity Degrading our farmland (int-3323)

4.2 INQUIRY ACTIVITIES

1. Use the following labels to create a flow diagram showing how the clearing of native vegetation can affect all four of the Earth's spheres.
   - Soil is left bare and exposed to wind and water erosion.
   - There is less evaporation of water from vegetation.
   - There is a loss of habitat for birds, animals and insects.
   - Increased water runs off from exposed land.
   - Increased sediment builds up in streams. **Classifying, organising, constructing**

2. (a) Rank what you consider to be the three most serious *environmental* issues shown on the **FIGURE 2** map.
   (b) What criteria did you use to make your decisions? (For example: extent of area covered, number of people affected, economic impacts.)
   (c) Share your list with at least two other students. What were the similarities and differences in the rankings?
   (d) Would you now alter your own ranking? Why or why not? **[Critical and Creative Thinking Capability]**

3. Select one agricultural product in Australia and conduct research to find data on how much is produced and how this has *changed* over time. **Examining, analysing, interpreting**

4.3 Changing our forest biome

4.3.1 Why are forests important?

In pre-industrial times, nearly 45 per cent of the world's land surface was covered in forest. Today, this figure is only 30 per cent. With industrialisation, technological development and population growth, large-scale **deforestation** has occurred as a result of the increasing need over time for timber products and land for food. It is estimated that of the forest cover lost, 85 per cent can be readily attributed to human activity — with 30 per cent due to clearing, 20 per cent through degradation and 35 per cent through fragmentation. Agricultural use now accounts for 37 per cent of the Earth's land surface.

Human society, the global economy and forests are interconnected, with more than one billion people depending on forests and forest products. Forest biomes offer us many goods and services, from providing wood and food products to supporting biological diversity. They provide habitat for a wide range of animals, plants and insects. Forests contribute to soil and water conservation, and they absorb **greenhouse gases**. Despite the growing awareness of the value of preserving forests, large-scale clearing continues. **FIGURE 1** shows the annual rate of deforestation in Brazil, while **FIGURE 2** shows the cumulative amount of forest lost over time.

4.3.2 Why do we clear forests?

By clearing forests, trees can be harvested for timber and paper production, and valuable ores and minerals can be mined from below the Earth's surface. Sometimes, forests are flooded rather than cleared in order to construct dams for hydroelectricity. Forests may also be cleared for food production, such as small-scale subsistence farming, large-scale cattle grazing, and for **plantations** and crop cultivation.

FIGURE 1 Annual loss of Amazon forest, Brazil, 1988–2018

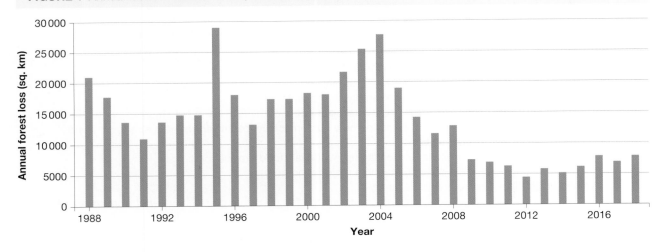

FIGURE 2 Total loss of Amazon forest, Brazil, since 1970

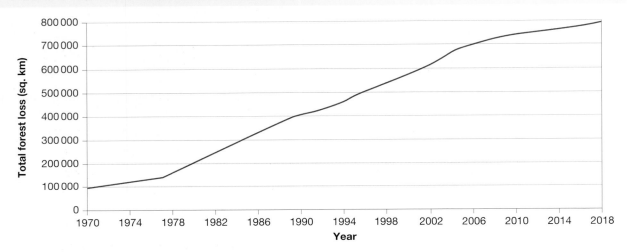

Road construction, usually funded by governments, also plays a part in changing rainforest environments (see **FIGURE 3**). Roads help to improve access and make more land available, especially to the landless poor. They also reduce population pressures elsewhere by encouraging people to move to new places. At the same time, businesses benefit from improved access to mining resources and forest timbers, and are better able to establish large cattle ranches and farms.

FIGURE 3 The effects of road building in the Amazon. Settlements tend to follow a linear pattern along the roads and then gradually move inland, opening up the forests.

4.3.3 What happens when forests are cleared?

FIGURE 4 illustrates some changes that forest clearing in the Amazon can have on the environment.

FIGURE 4 Impacts of clearing the Amazon forest

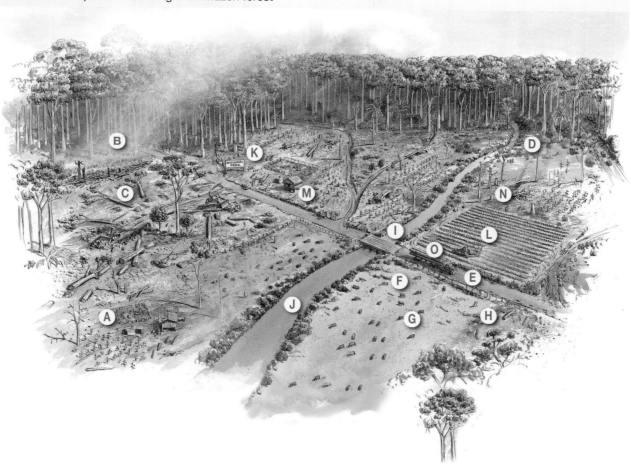

(A) New farmland with mixed crops established

(B) Smoke from clearing and burning

(C) Newly cleared land, trees cut down and burned. This is called slash-and-burn agriculture.

(D) Weeds and exotic species invade edges of remaining forest

(E) New road gives access to more settlers and to animal poachers

(F) Large cattle ranch

(G) Introduced cattle erode the fragile topsoil with their hard hooves.

(H) Erosion of topsoil increases, caused by rain on exposed soils.

(I) Flooding increases as the stream channel is clogged with sediment.

(J) The river carries more sediment as soil is washed into streams.

(K) Fences stop movement of rainforest animals in search of food.

(L) Pesticides and fertilisers wash into the river.

(M) Farm is abandoned as soil fertility is lost

(N) Weeds and other species dominate bare land.

(O) Harvesting of timber reduces forest biodiversity.

Explore more with my**World**Atlas

Deepen your understanding of this topic with related case studies and questions.
- Investigate additional topics > Environments > **Forest environments**

4.3 INQUIRY ACTIVITIES

1. Research soya-bean farming in the Amazon. How does it compare with cattle ranching in terms of **environmental sustainability**? **Comparing and contrasting**
2. Examine the illustration of rainforest destruction shown in **FIGURE 4**. Draw a sketch of what you predict the area will look like in ten years' time. Use labels and arrows to show important features. **Evaluating, predicting, proposing**

4.3 EXERCISES

Geographical skills key: GS1 Remembering and understanding **GS2** Describing and explaining **GS3** Comparing and contrasting **GS4** Classifying, organising, constructing **GS5** Examining, analysing, interpreting **GS6** Evaluating, predicting, proposing

4.3 Exercise 1: Check your understanding

1. **GS1** Refer to **FIGURE 2**. Describe the total loss in Brazilian forests since 1970. Use data in your answer.
2. **GS1** What are the advantages and disadvantages of road building in the Amazon?
3. **GS2** Why would subsistence farming in the Amazon be referred to as slash-and-burn farming?
4. **GS2** In what ways would the **environmental changes** of small-**scale** subsistence farming differ from those of large-**scale** soya-bean cropping?
5. **GS2** Why can the large-**scale** clearing of tropical rainforests be considered an **unsustainable** practice?

4.3 Exercise 2: Apply your understanding

1. **GS6** Opening up the rainforest with roads can lead to fragmentation of the forest. What might the effect of this be on:
   (a) native animals
   (b) local indigenous populations?
2. **GS3** Compare how a small-**scale** farmer from the Amazon and an **environmentalist** from another country might view the resources of a rainforest.
3. **GS5** Refer to **FIGURE 4**. Identify three **changes** to the river as a result of forest clearing.
4. **GS6** How might **changes** as a result of forest clearing affect farming downstream?
5. **GS2** Suggest two methods that could be used in the Amazon to reduce the amount of sediment washing into the river.

Try these questions in learnON for instant, corrective feedback. Go to www.jacplus.com.au.

4.4 SkillBuilder: GIS — deconstructing a map

What is GIS?

A geographical information system (GIS) is a storage system for information or data, which is stored as numbers, words or pictures.

Select your learnON format to access:

- an explanation of the skill (Tell me)
- a step-by-step process to develop the skill, with an example (Show me)
- an activity to allow you to practise the skill (Let me do it)
- questions to consolidate your understanding of the skill.

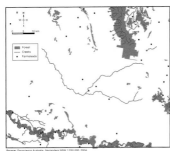

 Resources

Video eLesson GIS deconstructing a map (eles-1730)

Interactivity GIS deconstructing a map (int-3348)

4.5 Paper profits, global losses?

4.5.1 How do we use paper?

Look around you — how many different things can you see that are made out of paper? Paper is a renewable, recyclable material, and yet something so common and so useful also poses significant environmental consequences to biomes, at all stages of its usage cycle.

Biomes enable us to produce the food we eat, and they also supply many of the raw materials for manufacturing, such as minerals, ores and fibres. We are able to make cloth from cotton and wool using grassland biomes, while forest biomes give us wood for construction and **pulp** for making paper products. There are thousands of everyday items made from paper, ranging from toilet paper to disposable nappies, packaging, money, tickets and writing paper (see **FIGURE 1**).

Traditionally, paper has provided us with the means to record ideas, news, knowledge and even works of art. Paper is interconnected with social development as it aids in literacy and communication. Despite the advent of modern electronic communication, plastic bags and the 'paperless office', paper still remains an essential part of our homes and workplaces.

FIGURE 1 Different paper products

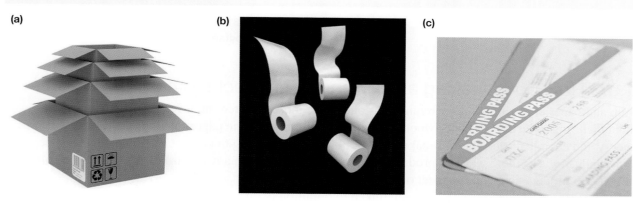

(a) (b) (c)

4.5.2 Are we really a paperless society?

We have become very dependent on paper products, with global consumption increasing by 400 per cent since the late 1970s. Today the world consumes 300 million tonnes of paper annually, and this is expected to rise as developing countries in Asia increase their consumption. China has now surpassed the United States as the largest producer and consumer of paper products.

The paper industry now consumes nearly 50 per cent of all industrial wood logged and is the fifth-largest consumer of energy needed in the production process. While the demand for some paper products has declined (for example, printed magazines and newspapers), the rapidly expanding e-commerce sector has seen a greater demand for packaging material as people purchase items online.

There is a strong interconnection between a country's level of economic development and its consumption of paper products. The use of tissue products, such as paper towels, facial tissues and toilet paper, is a good example of this relationship. As living standards within a country improve there is a changing attitude towards hygiene and improved lifestyle, which then encourages more spending on associated products. **FIGURE 2** shows that on a regional scale, North America, Australasia and Western Europe are the largest users of tissue products. Supporting this is the fact that in the United States, on average, 141 rolls of toilet paper are consumed per person per year, compared to Germany with 134 rolls, Japan with 91 and China with 49 rolls.

It was thought that with the introduction of personal computers we would become a paperless society, but this is far from the case. It has been estimated that in the United Kingdom, 45 per cent of paper in the office has a lifespan of less than one day. With technological advancements, paper has now become a cheap, disposable product. This has resulted in a high level of both usage and waste.

FIGURE 2 Regional consumption of paper tissue per capita, 2017

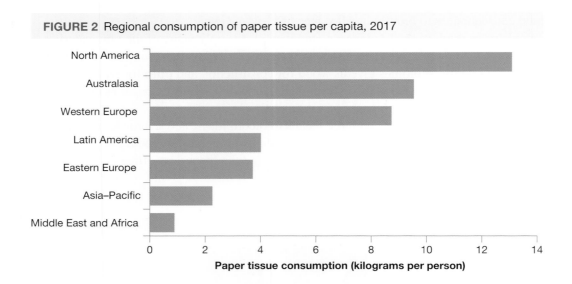

Paper tissue consumption (kilograms per person)

4.5.3 How does making and using paper affect biomes?

Pulp and paper production has been ranked as one of the most resource-intensive and highly polluting manufacturing industries. Besides wood fibre, the main inputs into the paper-making process are water, energy and chemicals needed for breaking down fibres and bleaching to create clean, white paper. The paper industry uses more water to produce one tonne of product than any other industry. It takes 10 litres of water to produce just one A4 sheet of paper.

Pulp and paper manufacturing has been linked to pollution of air, water and land through the discharge of toxic wastes. Paper's impact on biomes starts at the forest stage with the activity of logging the timber, and continues with the conversion of the timber to pulp and paper. Activities such as the logging of **old-growth forests** can have significant impacts on native animal populations and biodiversity. Environmental impacts continue even after paper has been used and discarded. In Australia, we waste on average 5.6 million tonnes of paper products per year, which equates to 229 kilograms per person. About 60 per cent is recycled and 40 per cent sent to landfill.

As paper decomposes, it releases methane, a dangerous greenhouse gas. **FIGURE 4** illustrates how biomes are affected by paper production.

FIGURE 3 Environmental effects of harvesting plantation timber

FIGURE 4 Environmental consequences of paper production

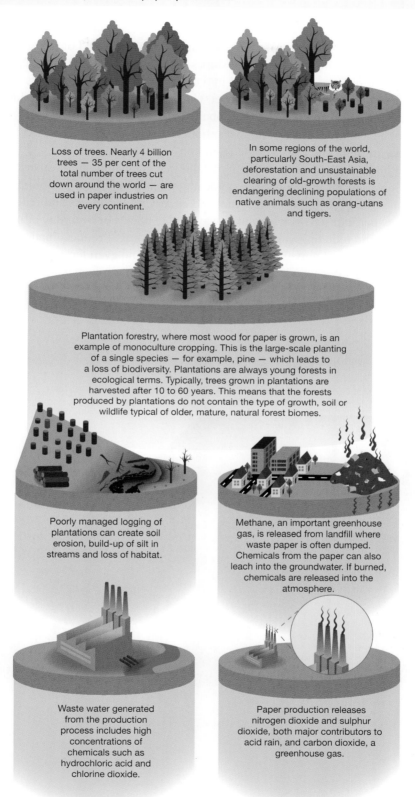

Loss of trees. Nearly 4 billion trees — 35 per cent of the total number of trees cut down around the world — are used in paper industries on every continent.

In some regions of the world, particularly South-East Asia, deforestation and unsustainable clearing of old-growth forests is endangering declining populations of native animals such as orang-utans and tigers.

Plantation forestry, where most wood for paper is grown, is an example of monoculture cropping. This is the large-scale planting of a single species — for example, pine — which leads to a loss of biodiversity. Plantations are always young forests in ecological terms. Typically, trees grown in plantations are harvested after 10 to 60 years. This means that the forests produced by plantations do not contain the type of growth, soil or wildlife typical of older, mature, natural forest biomes.

Poorly managed logging of plantations can create soil erosion, build-up of silt in streams and loss of habitat.

Methane, an important greenhouse gas, is released from landfill where waste paper is often dumped. Chemicals from the paper can also leach into the groundwater. If burned, chemicals are released into the atmosphere.

Waste water generated from the production process includes high concentrations of chemicals such as hydrochloric acid and chlorine dioxide.

Paper production releases nitrogen dioxide and sulphur dioxide, both major contributors to acid rain, and carbon dioxide, a greenhouse gas.

4.5.4 Is there a sustainable future for paper?

There has been an increase in environmental awareness and government legislation in recent years — this is helping to make the paper industry and people's use of paper more environmentally friendly, thus reducing the negative impacts on biomes. The Australian timber industry follows the Forest Code of Practice, which sets out rules and regulations regarding logging on slopes and protecting streams and habitat trees.

There has been considerable research conducted into the use of non-wood products, for example bamboo, sugarcane, hemp and **kenaf**, to provide fibre for paper. Currently these non-wood products make up only 10 per cent of the fibres used in paper production globally. Alternate fibres are critically important, however, in countries that do not have enough trees for paper production. Approximately 70 per cent of the raw materials used for paper in India and China come from non-wood products.

In some places, tree plantations are able to grow on land that is unsuitable for other forms of agriculture or is badly degraded, in which case there is likely to be an *increase* in habitat and biodiversity.

For the paper industry, the goals are to reduce fuel and energy requirements and reduce emissions. China, with its relatively new paper industry, is leading the way in this field. For everyday citizens, it is about making sensible choices, and reducing our use of paper and recycling. Products made from recycled paper can include masking tape, hospital gowns, bandages, egg cartons and even lampshades. However, paper can be recycled only 5 to 7 times, after which the fibres become too short and weak to bond together.

For every tonne of paper *not* consumed, the following savings are made:

- 18 trees
- 67 500 litres of water
- 9500 kilowatt hours of power
- 3300 kilograms of greenhouse gas emissions.

FIGURE 5 Paper recycling factory

DISCUSS

Do individuals have a responsibility to take action to reduce paper consumption or is this the responsibility of businesses and government? As a class, consider how you can reduce your paper consumption, as individuals, as a class and at the whole-school level. How could you encourage others to reduce consumption?

[Ethical Capability]

 Resources

🔗 **Weblink** Paper production

4.5 INQUIRY ACTIVITIES

1. Construct a list of the ten most important paper items in your day-to-day life. How does your list compare with those of others in your class? Are there similarities and differences? Why is this?

Comparing and contrasting

2. Use the **Paper production** weblink in the Resources tab to create a summary of key points for things you can do to promote *sustainable* paper production and consumption practices.

Examining, analysing, interpreting

4.5 EXERCISES

Geographical skills key: GS1 Remembering and understanding **GS2** Describing and explaining **GS3** Comparing and contrasting **GS4** Classifying, organising, constructing **GS5** Examining, analysing, interpreting **GS6** Evaluating, predicting, proposing

4.5 Exercise 1: Check your understanding

1. **GS2** What is *old-growth forest*? Why is it important to protect these types of biomes from being cleared?
2. **GS3** How might the structure and biodiversity be different in an old-growth forest compared to a tree plantation?
3. **GS2** Refer to **FIGURE 2**. With the use of data, describe the general consumption of tissue products globally.
4. **GS1** Refer to **FIGURE 3**. List the *environmental* effects of clearing a pine plantation that you can see in this photograph.
5. **GS2** What factors might contribute to a region increasing its consumption of tissue products in the future?

4.5 Exercise 2: Apply your understanding

1. **GS3** Refer to **FIGURE 2**. Suggest reasons why tissue consumption (paper towel, facial tissues, toilet paper etc.) has much higher consumption per capita in some regions than in others.
2. **GS2** How might a country's growing economy *change* its citizens' paper consumption?
3. **GS6** Do paperless societies exist? Do you think they might in the future? Write a paragraph expressing your viewpoint.
4. **GS6** Does your school have a paper recycling program? How effective do you think it is? How could it be improved? If there is not currently a program, outline measures you think could be taken to implement one.
5. **GS2** Many *places*, including Victoria, have often exported their waste (including paper products) to overseas countries for disposal. (China and the Philippines are now clamping down on this practice.) What is one advantage and one disadvantage of exporting paper waste?

Try these questions in learnON for instant, corrective feedback. Go to www.jacplus.com.au.

4.6 Depleting our bountiful ocean biome

4.6.1 Overfishing — causes and consequences

The ocean biome has always been seen as an unlimited resource of food for humans. In fact, overfishing is causing the collapse of many of our most important marine ecosystems, and threatens the main source of protein for over one billion people worldwide. **Aquaculture** is a possible solution but, at the same time, it contributes to the decline in fish stocks.

Overfishing is simply catching fish at a rate higher than the rate at which fish species can repopulate. It is an unsustainable use of our oceans and freshwater biomes.

Massive improvements in technology have enabled fish to be located and caught in larger numbers and from deeper, more inaccessible waters. The use of spotter planes, radar and factory ships ensures that fish can be caught, processed and frozen while still at sea.

Globally, fish is the most important animal protein consumed (see **FIGURE 1**). Historically, a lack of conservation and management of fisheries, combined with rising demand for fish products, has seen a 'boom and bust' mentality (see **FIGURE 2**). The larger fish species are targeted and exploited and, after their populations are decimated, the next species are fished. Examples of these include blue whales, Atlantic cod and bluefin tuna.

FIGURE 1 Global protein demand, 1980–2030 (million tonnes)

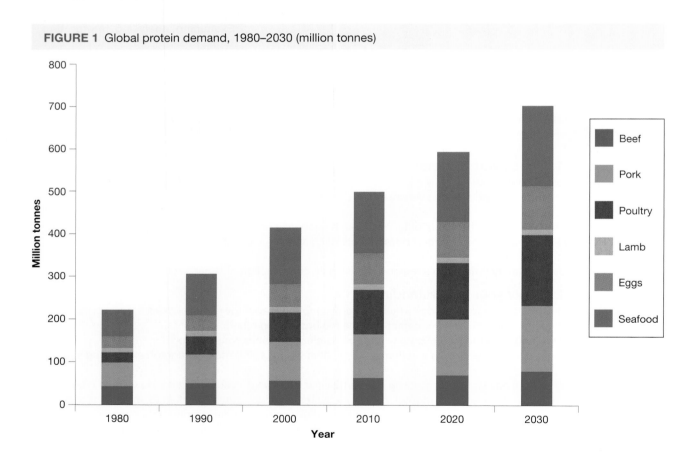

FIGURE 2 Unsustainable fishing

What happens when we overfish?

- With overfishing there are often large quantities of by-catch. This means that juvenile fish and other animals, such as dolphins and sea birds, are swept up in nets or baited on hooks before being killed and discarded. For every kilogram of shrimp caught in the wild, 5 kilograms of by-catch are wasted (see **FIGURE 3**).
- Destructive fishing practices such as cyanide poisoning, dynamiting of coral reefs and bottom trawling (which literally scrapes the ocean floor) cause continual destruction to local ecosystems.
- A large quantity of fish that could have been consumed by people is converted to fishmeal to feed the aquaculture industry, to fatten up pigs and chickens, and to feed pet cats (see **FIGURE 4**).
- Coastal habitats are under pressure. Coral reefs, mangrove wetlands and seagrass meadows, all critical habitats for fish breeding, are being reduced through coastal development, overfishing and pollution.

FIGURE 3 Up to 80 per cent of some fish catches is by-catch.

FIGURE 4 In Australia, the average cat eats 13.7 kilograms of fish a year. The average Australian eats 15 kilograms per year.

4.6.2 Is fish farming the solution?

Aquaculture is one of the fastest-growing food industries, providing fish for domestic and export markets. It brings economic benefits and increased food security (see **FIGURE 5**).

Since 2014, fish farming has produced more fish than fish caught in the wild; a harvest of 80 million metric tons was recorded in 2016. China is the largest farmed-fish producer, followed by India, Indonesia, Vietnam, Bangladesh, Egypt and Norway. Australia's history of fish farming started more than 6000 years ago, when Indigenous Australians created a series of fish traps in Lake Condah, in south-west Victoria, to capture a reliable source of eels (see subtopic 3.7). Today, aquaculture is Australia's fastest growing primary industry, producing more than 40 per cent of Australia's seafood.

FIGURE 5 Global fish and aquaculture 1990–2030

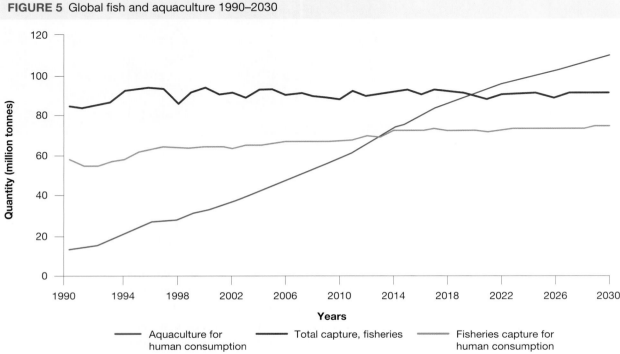

While aquaculture is often seen as a sustainable and eco-friendly solution to overfishing, its rapid growth and poor management in many places has created large-scale environmental change. Some of these changes are described below.

- *Pollution.* Many fish species are fed a diet of artificial food in dry pellets (see **FIGURE 6**). Chemicals in the feed, and the massive waste generated by fish farms, can pollute the surrounding waters.
- *Loss of fish stock.* Food pellets are usually made of fish meal and oils. Much of this comes from by-catch, but the issue is still that we are catching fish to feed fish. It can take 2 to 5 kilograms of wild fish to produce one kilogram of farmed salmon. Other ingredients in the food pellets include soybeans and peanut meal — products that are suitable for human consumption and grown on valuable farmland.
- *Loss of biodiversity.* Many of the fish species farmed are selectively bred to improve growth rates. If accidentally released into the wild, they can breed with native species and change their genetic makeup. This can lead to a loss of biodiversity. Capture of small ocean fish, such as anchovies, depletes food for wild fish and creates an imbalance in the food chain.
- *Loss of wetlands.* Possibly the greatest impact of aquaculture is in the loss of valuable coastal wetlands. In Asia, over 400 000 hectares of mangroves have been converted into shrimp farms. Coastal wetlands provide important ecological functions, such as protecting the shoreline from erosion and providing breeding grounds for native fish.

FIGURE 6 Feeding fish in pens, Thailand

 Resources

 Interactivity Hook, line and sinker (int-3324)

4.6 INQUIRY ACTIVITIES

1. Investigate and write a newspaper article on the collapse of the Atlantic cod fishery in Newfoundland. What lessons in the *sustainability* of fishing can be learned from the case of the Atlantic cod?

   Examining, analysing, interpreting

2. Collect photographs and other information to create an annotated poster showing one of the destructive fishing practices mentioned in this subtopic.

   Classifying, organising, constructing

4.6 EXERCISES

Geographical skills key: GS1 Remembering and understanding **GS2** Describing and explaining **GS3** Comparing and contrasting **GS4** Classifying, organising, constructing **GS5** Examining, analysing, interpreting **GS6** Evaluating, predicting, proposing

4.6 Exercise 1: Check your understanding

1. **GS2** Explain how overfishing can lead to a loss of biodiversity.
2. **GS1** What is aquaculture?
3. **GS1** What is by-catch?
4. **GS1** List three benefits and three drawbacks of fish farming.
5. **GS2** Why is it difficult to manage wild fish capture and prevent overfishing?

4.7 SkillBuilder: Interpreting a geographical cartoon

What are geographical cartoons?
Geographical cartoons are humorous or satirical drawings on topical geographical issues, social trends and events. A cartoon conveys the artist's perspective on a topic, generally simplifying the issue.

Select your learnON format to access:
- an explanation of the skill (Tell me)
- a step-by-step process to develop the skill, with an example (Show me)
- an activity to allow you to practise the skill (Let me do it)
- questions to consolidate your understanding of the skill.

 Resources

  Video eLesson Interpreting a geographical cartoon (eles-1731)

  Interactivity Interpreting a geographical cartoon (int-3349)

4.8 Losing the land

4.8.1 What is land degradation?

Land is one of our most basic resources and one that is often overlooked. In our quest to produce as much as possible from the same area of land, we have often failed to manage it sustainably. Land **degradation** is the result of such poor management.

Land degradation is a decline in the quality of the land to the point where it is no longer productive. Land degradation covers such things as soil **erosion**, invasive plants and animals, **salinity** and desertification. Degraded land is less able to produce crops, feed animals or renew native vegetation. There is also a loss in soil fertility because the top layers, rich in **humus**, can be easily eroded by wind or water. In Australia, it can take up to 1000 years to produce just three centimetres of soil, which can be lost in minutes in a dust storm.

Globally, 75 per cent of the Earth's land area is substantially degraded. The rate of fertile soil loss is now averaging 24 billion tons per year globally. In Australia, of the five million square kilometres of land used for agriculture, more than half has been affected by, or is in danger of, degradation.

Land degradation is common to both the developed and developing world, and results from both human and natural causes.

Human causes

Human causes of land degradation involve unsustainable land management practices, such as:

- *land clearance* — deforestation or excessive clearing of protective vegetation cover
- *overgrazing of animals* — plants are eaten down or totally removed, exposing bare soil, and hard-hoofed animals such as cows and sheep compact the soil (see **FIGURE 1**)
- *excessive irrigation* — can cause watertables to rise, bringing naturally occurring salts to the surface, which pollute the soil
- *introduction of exotic species* — animals such as rabbits and plants such as blackberries become the dominant species
- *decline in soil fertility* — caused by continual planting of a single crop over a large area, a practice known as monoculture
- *farming on marginal land* — takes place on areas such as steep slopes, which are unsuited to ordinary farming methods.

FIGURE 1 Soil erosion as a result of overgrazing in Australia

Biophysical causes

Natural processes such as prolonged drought can also lead to land degradation. However, land can sometimes recover after a drought period. Topography and the degree of slope can also influence soil erosion. A steep slope is more prone to erosion than flat land.

4.8.2 Impacts of land degradation

As land becomes degraded, productivity, or the amount of food it can produce, is lost. Some countries in sub-Saharan Africa have lost up to 40 per cent productivity in croplands over two decades, while population has doubled in the same time period. Farmers may choose to abandon the land, try to restore the land or, if the pressure to produce food is too great, they may have no choice but to continue using the land. Unproductive land will be exposed to continual erosion or weed invasion.

If extra fertilisers are applied to try to improve fertility, the excessive nutrients can create pollution and algae build-up in nearby streams. Airborne dust creates further hazards for both people and air travel. Land degradation is a classic example of human impact on all spheres of the environment — atmosphere, biosphere, lithosphere and hydrosphere.

FIGURE 2 illustrates global land productivity for different land uses, which is an important indicator of land degradation. Twenty per cent of the world's cropland shows declining or stressed land productivity, despite the efforts and resources being used to maintain food production.

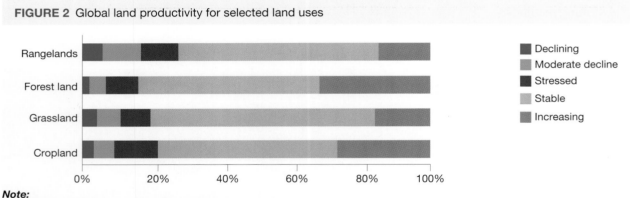

FIGURE 2 Global land productivity for selected land uses

Legend:
- Declining
- Moderate decline
- Stressed
- Stable
- Increasing

Note:
Rangelands refers to shrublands mostly used for grazing.
Forest land applies to land with more than 40 per cent tree cover.
Grassland includes natural grasslands and pasture for grazing.
Cropland includes all arable land and where 50 per cent of land is used for crops.

About 40 per cent of degraded lands are found in places that experience widespread poverty, which is a contributing factor to food insecurity. Poor farmers with degraded land and few resources often have little choice but to continue to work the land. There is a strong interconnection between land degradation, migration and political instability. If declining soil quality and an increase in droughts due to climate change continue, between 50 and 700 million people could be forced to move by 2050.

Desertification is an extreme form of land degradation. It usually occurs in semi-arid regions of the world, and the result gives the appearance of spreading deserts. Desert biomes, or arid regions, are harsh, dry environments where few people live. In contrast, semi-arid regions, or drylands, occupy 41 per cent of the Earth's surface and support over two billion people, 90 per cent of whom live in developing nations. Economically, drylands support 44 per cent of the world's food production and 50 per cent of the world's livestock. Although traditional grazing and cropping has taken place in dryland regions for centuries, population growth and the demand for food have put enormous pressure on land resources. Overclearing of vegetation, overgrazing and overcultivation are a recipe for desertification.

on Resources

Interactivity Losing land (int-3325)

4.8 INQUIRY ACTIVITIES

1. Create an annotated sketch to show the **interconnection** between plants and soil. Use the following points as labels on your sketch.
   - Plant roots help hold soil together.
   - Decomposing plants add nutrients to the soil.
   - Plants shade the topsoil and reduce evaporation.
   - Plants reduce the speed of wind passing over the ground.   **Classifying, organising, constructing**
2. Investigate an area in Victoria that is suffering from land degradation. Identify the location, causes and impacts of the degradation. Are any steps being taken to reduce the impacts?

Examining, analysing, interpreting

4.8 EXERCISES

Geographical skills key: GS1 Remembering and understanding **GS2** Describing and explaining **GS3** Comparing and contrasting **GS4** Classifying, organising, constructing **GS5** Examining, analysing, interpreting **GS6** Evaluating, predicting, proposing

4.8 Exercise 1: Check your understanding

1. **GS1** List four human causes of land degradation.
2. **GS1** List two natural causes of land degradation.
3. **GS1** Which biome supports more life: desert or drylands? Why?
4. **GS2** Explain how land degraded by drought may recover, whereas land degraded by cultivation may not.
5. **GS2** Consider the photograph in **FIGURE 1**. Why would it be difficult to either graze animals or grow crops on this land?

4.8 Exercise 2: Apply your understanding

1. **GS6** Examine the photograph in **FIGURE 1**. If this was your property and your livelihood, what steps would you take to reduce the erosion problem?
2. **GS5** Refer to **FIGURE 2**.
   (a) Which land cover has the greatest percentage of stressed and declining productivity?
   (b) What type of farming activities could explain the increased productivity in croplands?
3. **GS2** Why is poverty often linked to food insecurity?
4. **GS6** How might climate **change** in the future contribute to desertification in semi-arid regions of the world?
5. **GS6** Propose three steps that a farmer could take to reduce the impact of overgrazing.

Try these questions in learnON for instant, corrective feedback. Go to www.jacplus.com.au.

4.9 The effects of farmland irrigation

4.9.1 What is the purpose of irrigation?

Food production and security is directly related to water availability. Water is a finite resource and, although there is plenty of water in the world, it is not always located where people are concentrated or where food is grown. Therefore, humans have drawn water from both surface and underground sources to improve food production in areas of high population.

Most of the world's food production is rain fed, or dependent on naturally occurring rainfall. Only a small proportion of agricultural land is irrigated, yet **irrigation** is now the biggest user of water in the world, consuming 70 per cent of the world's freshwater resources. Irrigation brings many benefits, such as:

- supplementing or replacing rain, especially in places where rainfall is low or unreliable. In many parts of the world, it is not possible to produce food without irrigation.
- increasing crop yields, up to three times higher than rain-fed crops. Only 20 per cent of the world's farmland is irrigated but it produces over 40 per cent of our food.

- enabling a wide variety of foods to be grown, especially those with high water needs, such as rice, or with high value, such as fruit and wine grapes
- flexibility, being used at different times according to crop needs; for example, during planting and growing or close to harvest time.

FIGURE 1 Irrigation allows for pasture to be grown in times of drought. Compare the irrigated with the non-irrigated paddocks.

4.9.2 Environmental impacts of irrigation

While irrigation has resulted in increased food production and greater food security, it has also created major changes to the biomes where it is used. Irrigation changes the natural environment by extracting water from rivers and lakes and through the building of structures to store, transfer and dispose of water. The topography, or shape of the land, is often changed too, such as when terraces are built for paddy fields. In addition, irrigation water is often applied to the land in much larger quantities than naturally occurs, which can lead to changes in soil composition, and **waterlogging** and salinity problems.

How does irrigation create salinity problems?

Overwatering of shallow-rooted crops adds excess water to the **watertable**, causing it to rise (see **FIGURE 2**).

If the subsoils are naturally salty, much of this salt can be drawn to the surface. Most crops and pasture will not grow in salty soils, so the land becomes useless for farming. Land that is affected by salinity is also more prone to wind and water erosion.

FIGURE 2 The development of irrigation salinity

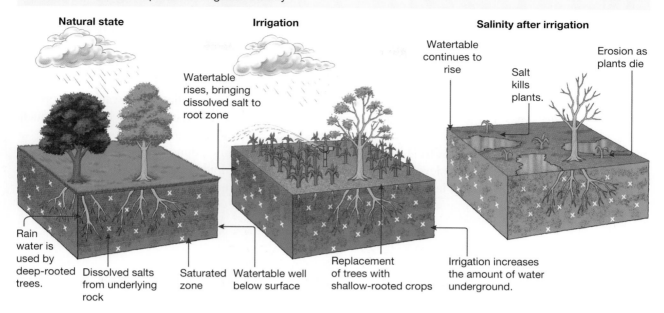

Globally, some 62 million hectares of land (an area the size of France) has been lost due to such issues. Salinity is also a major cause of land degradation in Australia (see **FIGURE 3**).

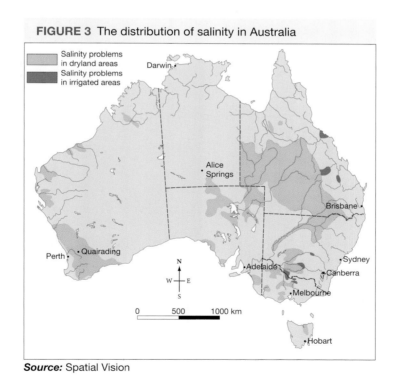

FIGURE 3 The distribution of salinity in Australia

Source: Spatial Vision

4.9.3 Impacts of diverting and extracting water

As population increases, so too does demand for water. Moreover, there are always competing demands for water from the domestic, industrial and environmental sectors. For countries that have growing populations and limited water resources, water deficits and food insecurity are a growing concern. In many places in the world, water is becoming increasingly scarce. Consequently, the development of water resources is becoming more expensive and, in some cases, environmentally destructive.

For thousands of years, farmers have diverted water from rivers, lakes and wetlands for watering crops and pastures in dry areas. Large-scale irrigation schemes can effectively 'water' our deserts but, if too much water is used, wetlands can dry out, rivers cease to flow and lakes and underground **aquifers** shrink. It is estimated that between three and six times more water is held in reservoirs around the world than exists in natural rivers. It is possible that the level of water extraction will nearly double by 2050.

Lake Chapala, Mexico's largest lake (see **FIGURE 4**), is shrinking. The amount of water lost through irrigation and domestic use, combined with high evaporation rates, has seen the volume of the lake decrease by 50 per cent.

FIGURE 4 Map of Lake Chapala, Mexico. Note the area of land drained for farmland.

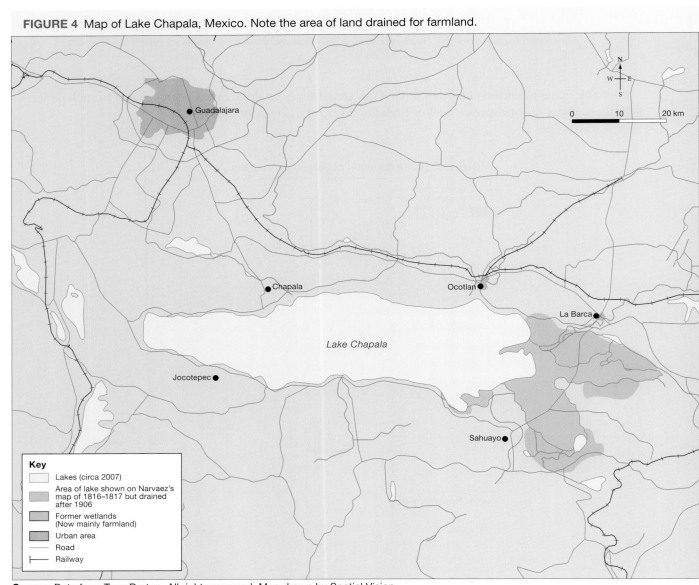

Source: Data from Tony Burton. All rights reserved. Map drawn by Spatial Vision.

As surface water resources become fully exploited, people turn to underground water sources. Improvements in technology have also enabled farmers to pump water from aquifers deep underground (see **FIGURE 5**).

FIGURE 5 Diagram showing the use of groundwater as a water source for farming

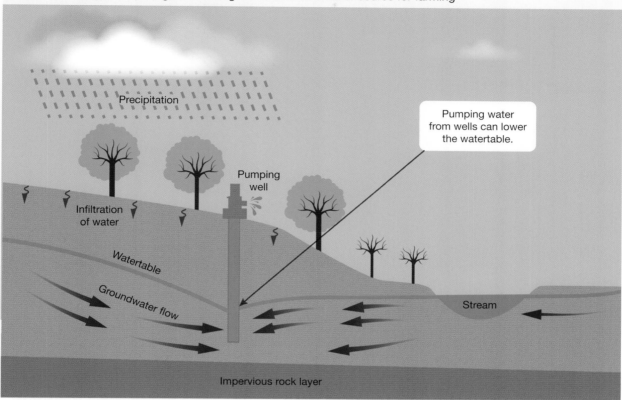

Groundwater levels do not respond to changes in the weather as rapidly as rivers and lakes do. If the water is removed unsustainably (at a rate that is faster than the rate of replenishment by rainfall, run-off or underground flow), then watertables fall. Water extraction then becomes harder and more expensive. Water stored in aquifers can take thousands of years to replenish. Over-extraction of groundwater can result in wells running dry, reduced stream flow, and even land subsidence (sinking).

The High Plains region of the central United States is the leading irrigation area in the western hemisphere, producing over $20 billion worth of food and fibre per year (see **FIGURE 6**). In all, 5.5 million hectares of semi-arid land is irrigated using water pumped from the huge Ogallala Aquifer (see **FIGURE 7**). Since large-scale irrigation was developed in the 1940s, groundwater levels have dropped by more than 30 metres. Pesticides

FIGURE 6 Irrigated cropland relies heavily on water from the Ogallala Aquifer.

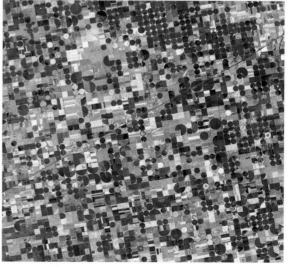

and other pollutants from farming have also infiltrated the groundwater. Scientists estimate that the aquifer will be 69 per cent depleted by 2060 and it would take more than 6000 years for it to refill naturally.

FIGURE 7 The size of the Ogallala Aquifer in the central United States

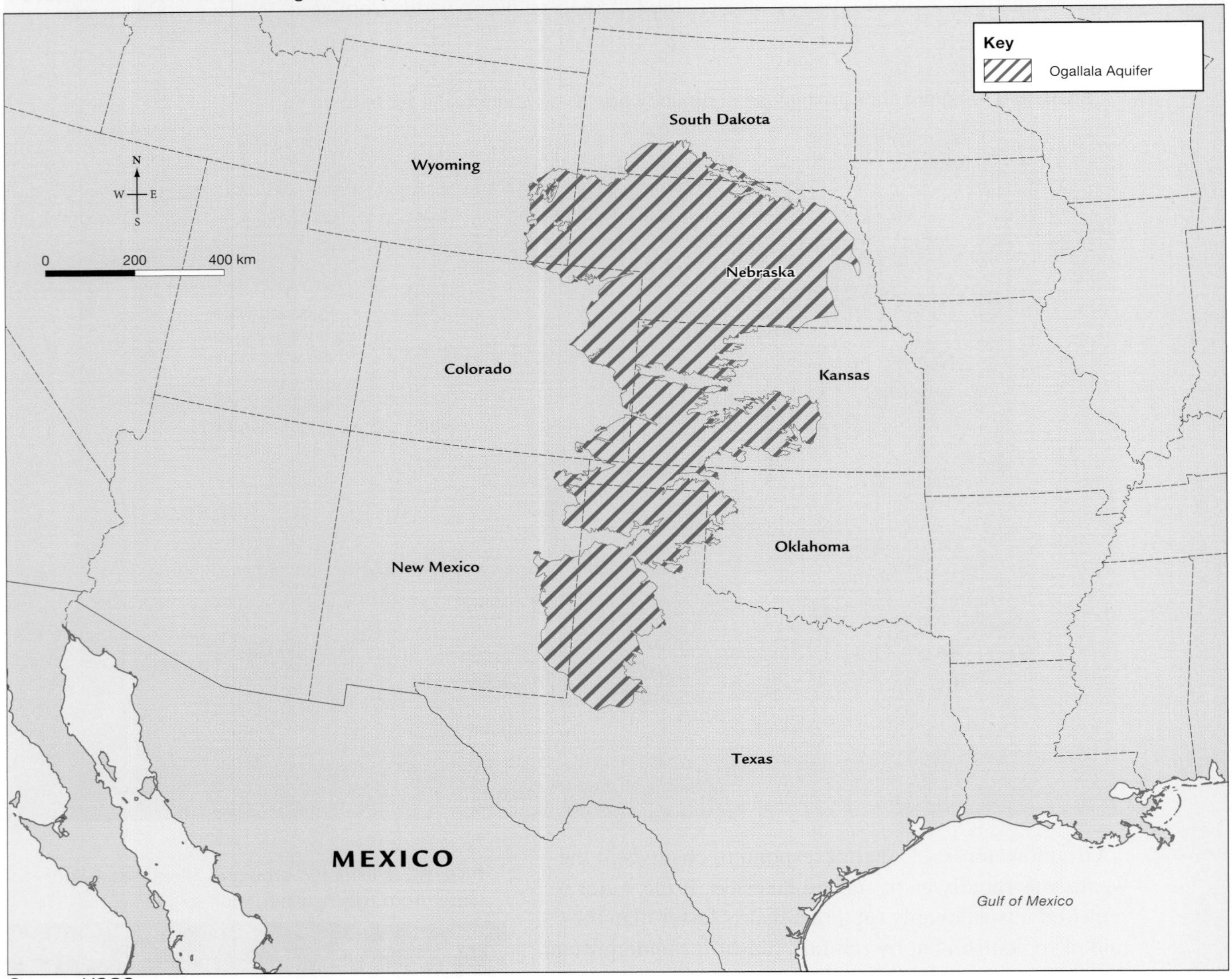

Key

Ogallala Aquifer

Source: USGS

4.9 INQUIRY ACTIVITIES

1. (a) Investigate methods used in Australia to reduce the ***environmental*** effects of salinity.
   (b) Using **FIGURE 2** as a model, create a similar sketch depicting the development of irrigation salinity. Based on your research findings, annotate your drawing with suggestions for how to reduce the effects of irrigation salinity.   **Classifying, organising, constructing**
2. Research the Aral Sea in Asia and write a report outlining:
   - location
   - the issue of over-extraction of water
   - impacts of overuse.

   Include a location map and labelled photographs in your report.   **Examining, analysing, interpreting**

4.9 Exercise 1: Check your understanding

1. **GS1** List the different types of water resources that can be used to supply water for food production.
2. **GS1** What is meant by the term *waterlogging*?
3. **GS1** What percentage of the world's fresh water is consumed by irrigation?
4. **GS2** What **changes** to the **environment** are needed in order to irrigate a large region?
5. **GS1** Apart from irrigation, what would be the other main uses of water?
6. **GS2** Study the **FIGURE 3** map, which shows the distribution of salinity in Australia.
   (a) Estimate the approximate percentage of each state affected by salinity.
   (b) Why do you think dryland salinity covers a larger area than irrigation salinity?
7. **GS2** Study **FIGURE 5**. Explain how pumping groundwater can lower watertables.
8. **GS3** Compare the advantages and disadvantages of using groundwater and surface water for farming.
9. **GS1** Identify one natural and one human factor that have contributed to the **change** in water levels in Lake Chapala.

4.9 Exercise 2: Apply your understanding

1. **GS5** Soil salinity was not a problem when Indigenous Australian peoples were the land's sole caretakers. What does this suggest about land management practices in this country since 1788?
2. **GS6** Has irrigation been a success or failure? Write a paragraph expressing your viewpoint.
3. **GS5** Refer to **FIGURE 7**. With the use of the *scale* bar, work out the approximate area covered by the Ogallala Aquifer.
4. **GS6** If the Ogallala Aquifer was to run dangerously low, and irrigation was no longer possible, what would be the short-term and long-term consequences?
5. **GS6** What are some of the likely effects of draining wetlands for farmland, as has occurred around Lake Chapala, Mexico?
6. **GS6** Select one of the examples shown in this subtopic and consider what steps water managers could take to reduce the impact of **unsustainable** water use in the region.

Try these questions in learnON for instant, corrective feedback. Go to www.jacplus.com.au.

4.10 Diminishing global biodiversity

4.10.1 The loss of biodiversity

The last few centuries have seen the greatest rate of species extinction in the history of the planet (see **FIGURE 1**). The population of most species is decreasing, and genetic diversity is declining, especially among species that are cultivated for human use. Six of the world's most important land biomes have now had more than 50 per cent of their area converted to agriculture (see **FIGURE 2**).

In those places where there has been very little industrial-scale farming, a huge variety of crops are still grown. In Peru, for example, over 3000 different potatoes are still cultivated. Elsewhere, biodiversity as well as agricultural biodiversity (biodiversity that is specifically related to food items) is in decline. In Europe, 50 per cent of all breeds of domestic animals have become extinct, and in the United States, 6000 of the original 7000 varieties of apple no longer exist. How has this happened?

- Industrial-scale farming and new high-yielding, genetically uniform crops replace thousands of different traditional species. Two new rice varieties in the Philippines account for 98 per cent of cropland.
- Converting natural habitats to cropland and other uses replaces systems that are rich in biodiversity with monoculture systems that are poor in diversity (see **FIGURE** 3).
- Uniform crops are vulnerable to pests and diseases, which then require large inputs of chemicals that ultimately pollute the soil and water. Traditional ecosystems have many natural enemies that combat pest species.
- The introduction of modern breeds of animals has displaced indigenous breeds. In the space of 30 years, India has lost 50 per cent of its native goat breeds, 30 per cent of sheep breeds and 20 per cent of indigenous cattle breeds.

FIGURE 1 Extinctions per thousand species per millennium

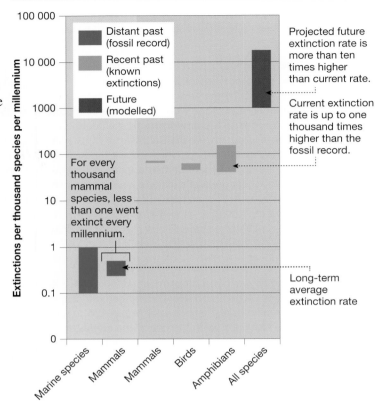

FIGURE 2 Percentage of biomes converted to agriculture over time

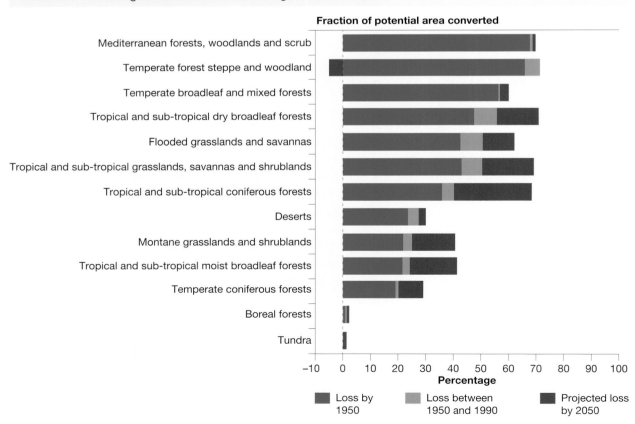

FIGURE 3 Changes to percentage of original species according to changes in biomes for food production

GRASSLAND

100%

Original species

Extensive use

Burning

Subsistence agriculture

Intensive agriculture

Abundance of original species

0%

4.10.2 Australia's biodiversity

Australia has a high number of **endemic** species, and 7 per cent of the world's total species of plants, animals and micro-organisms. This makes Australia one of only 17 countries in the world that are classified as megadiverse — having high levels of biodiversity. These 17 nations combined contain 75 per cent of the Earth's total biodiversity (see **FIGURE 4**). Australia's unique biodiversity is due to its 140 million years of geographic isolation. However, Australia has experienced the largest documented decline in biodiversity of any continent over the past 200 years. It is thought that 48 plant species and 50 species of animals (27 mammal species and 23 bird species) are now extinct.

FIGURE 4 Distribution of megadiverse countries

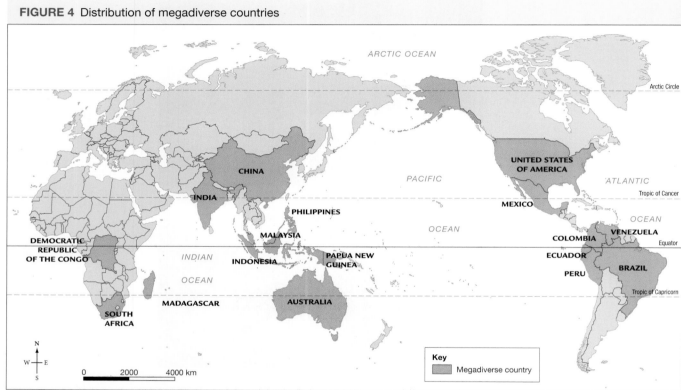

Source: Spatial Vision

4.10 INQUIRY ACTIVITY

Investigate the issue of whaling and the conflicting viewpoints held by Australia and Japan.
a. What were the factors (reasons) involved in:
  (i) Australia's decision to ban whaling
  (ii) Japan's decision to continue whaling?
b. How would you suggest the two countries could come to a resolution? **Examining, analysing, interpreting [Ethical Capability]**

4.11 Does farming cause global warming?

4.11.1 The connection of farming to global warming

Agriculture and climate change are interconnected processes, both of which take place on a global scale. Global warming will cause changes to what can be grown and where. At the same time, however, many of our agricultural practices may in fact contribute to global warming.

The term 'global warming' generally refers to the warming of the planet that is caused by increased emissions of **greenhouse gases** as a result of human activity. The human activity of food production can contribute to global warming in two ways.

- Grazing animals and flooded rice paddies produce the greenhouse gas methane. Livestock are thought to be responsible for 35 per cent of the world's methane output. The next largest sources, in order, are oil and gas, landfill, rice paddies and wastewater treatment systems.
- Food production changes the surface of the Earth, which then alters the planet's ability to absorb or reflect heat and light. Large-scale deforestation and desertification can significantly alter a region's microclimate. Around 80 per cent of global deforestation is caused by clearing the land for grazing, slash-and-burn farming and cropping.

4.11.2 The biggest polluters

Cows emit large quantities of methane through belching and flatulence (caused by digestive gases). The gas is produced by bacteria digesting grass in one of the four stomachs that cows have. It has been estimated that one cow could produce somewhere between 100 and 500 litres of methane per day (see **FIGURE 1**). This amount is similar to the pollution produced by one car in one day. When you consider there are over 1.5 billion cows in the world, this equates to a lot of gas. Scientists today are working on 'fuel-efficient cows' — cows that convert feed more efficiently into milk rather than methane.

FIGURE 1 Argentine scientists have strapped plastic tanks to the backs of cows to assess how much methane they produce.

FIGURE 2 A cartoonist's view of livestock and global warming

Rice farming is one of the biggest sources of human-produced methane, averaging between 50 and 100 million tonnes per year. The gas is produced in the warm, waterlogged soils of the rice paddies (see **FIGURE 3**).

FIGURE 3 Methane is released from rice paddies.

The practice of **factory farming**, in which a very high number of animals are concentrated in the one place, produces an unmanageable amount of waste (see **FIGURE 4**). On a sustainable farm, animal manure can be used as a natural fertiliser, but on a factory farm the large quantity becomes a source of methane, because the waste is often mixed with water and stored in large ponds or lagoons. An additional problem can occur if these ponds leak, as they create soil and water pollution. The use of nitrogen-based fertilisers on farms also releases nitrous oxide, another greenhouse gas.

FIGURE 4 Factory farming produces large quantities of waste products.

4.11.3 Deforestation's connection to global warming

Trees are 50 per cent carbon, so when they are burned or felled to create land for farming, the CO_2 they store is released back into the atmosphere. Research suggests that the loss of tropical rainforests alone accounts for 8 per cent of global carbon dioxide emissions, similar to those of the USA. Forests also act as carbon sinks, the most effective way of storing carbon. Large areas of cleared land absorb more heat than native vegetation, which can lead to changes in local weather conditions.

4.11 INQUIRY ACTIVITIES

1. Research the sources of methane gas and find out what percentage each contributes to world methane output. Present your information in a pie or bar graph. Is the biggest source natural or human?
   Examining, analysing, interpreting

2. If one cow produces the equivalent of 16 kWh of energy per day, how many cows would be needed to power your own home per day? You will need to check your household electricity bill.
   Examining, analysing, interpreting

3. How might climate *change* affect agriculture? Do some online research to investigate some of the possible effects of climate *change* on food production.   **Examining, analysing, interpreting**

4.11 EXERCISES

Geographical skills key: GS1 Remembering and understanding **GS2** Describing and explaining **GS3** Comparing and contrasting **GS4** Classifying, organising, constructing **GS5** Examining, analysing, interpreting **GS6** Evaluating, predicting, proposing

4.11 Exercise 1: Check your understanding

1. **GS1** What are three greenhouse gases?
2. **GS1** What are two ways in which deforestation can contribute to *changes* in greenhouse gases?
3. **GS1** What is the biggest global emitter of methane: livestock, cars or rice paddies?
4. **GS2** Is factory farming a *sustainable* form of food production? Give reasons for your answer.
5. **GS2** The building of large-*scale* dams and subsequent flooding of forests in the Amazon is also contributing to increases in greenhouse gas emissions. What is the reason for this?

4.11 Exercise 2: Apply your understanding

1. **GS2** Why do we have factory farming?
2. **GS3** Study **FIGURE 4**.
   (a) How does factory farming differ from traditional farming methods? Create a table with two columns, one headed 'Key features of a traditional dairy farm' and the other 'Key features of a factory farm'. List the features of both styles of farming and then compare your lists.
   (b) Write a paragraph summarising the similarities and differences between the two methods.
3. **GS6** Suggest some ideas for reducing agriculture's contribution to global warming.
4. **GS6** Agriculture and climate *change* are *interconnected* processes. Suggest possible ways that increased temperatures and increased frequency of storms could impact on food production in Australia.
5. **GS2** Explain how deforestation can *change* the microclimate, especially temperatures and moisture levels, of a region. (*Note:* A microclimate is a local set of atmospheric conditions that differ from those in the surrounding areas.)

Try these questions in learnON for instant, corrective feedback. Go to www.jacplus.com.au.

4.12 Thinking Big research project: Fished out! PowerPoint

SCENARIO

For an upcoming 'Biomes and food security' conference, you have been invited to give a presentation on a current issue relating to food security — in this case, overfishing — and to outline some of the responses that consider economic, social and environmental factors. You will need to conduct some background research and then produce a PowerPoint presentation to highlight the threat that overfishing represents to the world's food security.

Select your learnON format to access:
- the full project scenario
- details of the project task
- resources to guide your project work
- an assessment rubric.

 Resources

> **ProjectsPLUS** Thinking Big research project: Fished out! PowerPoint (pro-0190)

4.13 Review

4.13.1 Key knowledge summary
Use this dot point summary to review the content covered in this topic.

4.13.2 Reflection
Reflect on your learning using the activities and resources provided.

 Resources

> **eWorkbook** Reflection (doc-31718)
>
> Crossword (doc-31719)
>
> **Interactivity** The impacts of global food production crossword (int-7646)

KEY TERMS
aquaculture the farming of aquatic plants and aquatic animals such as fish, crustaceans and molluscs
aquifer a body of permeable rock below the Earth's surface, which contains water, known as groundwater
biophysical environment the natural environment, made up of the Earth's four spheres — the atmosphere, biosphere, lithosphere and hydrosphere
deforestation clearing forests to make way for housing or agricultural development
degradation deterioration in the quality of land and water resources caused by excessive exploitation

endemic describes species that occur naturally in only one region

erosion the wearing down of rocks and soils on the Earth's surface by the action of water, ice, wind, waves, glaciers and other processes

factory farming the raising of livestock in confinement, in large numbers, for profit

greenhouse gases any of the gases that absorb solar radiation and are responsible for the greenhouse effect. These include water vapour, carbon dioxide, methane, nitrous oxide and various fluorinated gases.

groundwater water that exists in pores and spaces in the Earth's rock layers, usually from rainfall slowly filtering through over a long period of time

humus an organic substance in the soil that is formed by the decomposition of leaves and other plant and animal material

irrigation the supply of water by artificial means to agricultural areas

kenaf a plant in the hibiscus family that has long fibres; useful for making paper, rope and coarse cloth

old-growth forests natural forests that have developed over a long period of time, generally at least 120 years, and have had minimal unnatural disturbance such as logging or clearing

plantation an area in which trees or other large crops have been planted for commercial purposes

pulp the fibrous material extracted from wood or other plant material to be used for making paper

salinity the presence of salt on the surface of the land, in soil or rocks, or dissolved in rivers and groundwater

waterlogging saturation of the soil with groundwater such that it hinders plant growth

watertable the surface of the groundwater, below which all pores in the soils and rock layers are saturated with water

5 Challenges to food security

5.1 Overview

The world produces enough food to feed everyone. So why do hundreds of millions of people go hungry every day?

5.1.1 Introduction

For these children, in a tent camp for people displaced by flooding in northern India, the only kind of food security comes in the form of aid. One in nine people in the world, or around 850 million, will go to bed hungry tonight. What is preventing everyone getting enough to eat? And if this is the current situation, what will happen in the future, with our population set to rise to nearly 10 billion by 2050? How can we ensure food security for all the people of our ever-growing world population?

on Resources

☑ **eWorkbook**   Customisable worksheets for this topic

🎞 **Video eLesson**   Food for thought (eles-1720)

LEARNING SEQUENCE

To access a pre-test and starter questions and receive immediate, **corrective feedback** and **sample responses** to every question, select your learnON format at www.jacplus.com.au.

5.2 Global food security

5.2.1 What is food security?

Very few Australians, by choice, would go to bed at night hungry. We live in a country where there is a plentiful supply and wide range of food items available. Our relatively high standard of living enables most of us to afford to purchase, store and prepare food, or even dine out. Most of us are secure in the knowledge that there will be food available at the next mealtime.

According to the United Nations Food and Agriculture Organization, 'Food security exists when all people, at all times, have physical and economic access to enough safe and nutritious food to meet their dietary needs and food preferences for an active and healthy lifestyle.'

Food security for you, as a student, means that your family either grows its own food, has sufficient income to purchase food, or is able to barter or swap food. Similarly, food security for a country means that it is able to grow sufficient food, or it has enough wealth to import food, or it combines the two. Not all people in the world are able to achieve this. Further, access to a wide variety of foods varies from place to place. For example, consider the range of foods available in the two markets in **FIGURES 1(a)** and **(b)**.

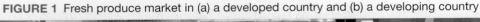

FIGURE 1 Fresh produce market in (a) a developed country and (b) a developing country

5.2.2 Who has food security?

The **FIGURE 2** map shows the countries of the world, scored according to the Global Food Security Index. This is based on a range of 12 different **indicators**, including the:

- affordability of food
- accessibility of food
- nutritional value of food
- safety of food
- nutritional and health status of the population.

Countries that have a high rating on the index are able to produce more food than they require, so they can export their surplus, or they are able to afford to import all of their food needs, as is the case for Singapore.

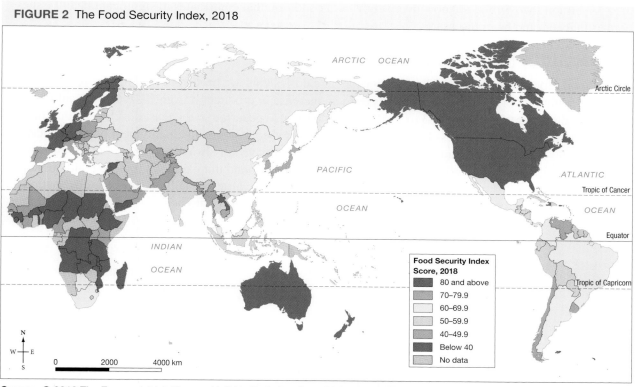

FIGURE 2 The Food Security Index, 2018

Food Security Index Score, 2018
- 80 and above
- 70–79.9
- 60–69.9
- 50–59.9
- 40–49.9
- Below 40
- No data

Source: © 2019 The Economist Intelligence Unit Limited, data from Global Food Security Index. Map drawn by Spatial Vision.

In Australia, we produce three times as much food as we consume. We are a major exporter of both fresh and processed food, and can trade competitively in cereals, oil seeds, beef, lamb, sugar and dairy products. Ninety per cent of our food is grown here in Australia. Of the remaining 10 per cent that we import, many foods are either processed or out of season in Australia; oranges are an example. Global trade is an important component of food security because it is almost impossible to exactly match food production to food demands.

As a country, Australia does not have a problem feeding its population, but it has a humanitarian interest in the food security of developing nations. As a major food producer, Australia does face future challenges. There is declining growth in agricultural productivity, the threat of climate change, and increasing competition for land and water.

5.2.3 Who is at risk of food insecurity?

FIGURE 2 also shows those countries that have a low Food Security Index score. It is estimated that more than 850 million people — one in every nine people in the world — are **undernourished**, with diets that are minimal or below the level of sustenance. Poor diet and limited access to food create large-scale food insecurity in many parts of Africa and southern Asia. People who do not have a regular and healthy diet often have shortened life expectancy and an increased risk of disease. Children are especially vulnerable to poor diet, and their growth, weight, and physical and mental development suffer. India is home to 24 per cent of the world's **malnourished** and 30 per cent (46.6 million) of the world's children under five with stunted growth due to poor and inadequate diets.

Paradoxically there is also an interconnection between food insecurity and obesity. When fresh food is scarce or expensive, people will choose cheaper food that is often high in kilojoules but low in nutrients. This is quite common in urban areas of middle- and high-income countries. Of the world's population of over 7.7 billion, two billion are now overweight — a condition that contributes to significant health issues such as diabetes and heart disease.

5.2.4 Why is there food insecurity?

Global food production now provides one-third more calories than are needed to feed the entire world. Since the beginning of this century there has been an increase in production from 2716 to 2904 calories per person per day. Increases from 2083 to 2358 calories have also occurred in the least developed countries. There is, however, unequal access to **arable** land, technology, education and employment opportunities. Improvements in food production and economic development have not always occurred in those places experiencing rapid growth in population. Food is redistributed around the world via trade and aid but neither is a long-term or large-scale solution to food insecurity. Regional variations still occur in the distribution of hunger, as can be seen in **FIGURE 3**. Since 2014, severe food insecurity has actually risen in Africa, Latin America, and the world as a whole.

FIGURE 3 Regional food insecurity, 2014–2017

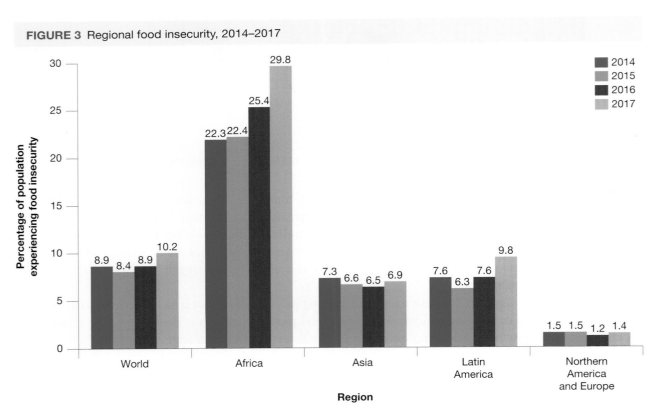

Some of the reasons for food insecurity include:

- poverty
- population growth
- weak economy and/or political systems
- conflict
- natural disasters such as drought.

5.2 INQUIRY ACTIVITIES

1. Research and find out the causes and effects of one of the conditions caused by dietary deficiency, such as deficiency in iron, vitamin A or vitamin C. **Examining, analysing, interpreting**
2. Select one of the **places** mapped in **FIGURE 2** as being at extreme risk of food insecurity. Find out the main factors that contribute to its food insecurity. **Examining, analysing, interpreting**
3. **GS2** Refer to a map of conflict in your atlas or online. Is there an **interconnection** between those countries that have a high or extreme risk of food insecurity and those countries that are experiencing conflict? Include country names in your answer. **Examining, analysing, interpreting**

5.2 EXERCISES

Geographical skills key: GS1 Remembering and understanding **GS2** Describing and explaining **GS3** Comparing and contrasting **GS4** Classifying, organising, constructing **GS5** Examining, analysing, interpreting **GS6** Evaluating, predicting, proposing

5.2 Exercise 1: Check your understanding

1. **GS1** What is meant by the term *food security*?
2. **GS3** Compare the two photographs in **FIGURES 1(a)** and **1(b)**.
   (a) What are the similarities and differences between the two markets?
   (b) Do you think all food groups would be available in both markets? Why or why not?
3. **GS4** Refer to **FIGURE 2**.
   (a) List five examples of countries, from different regions of the world, that are considered to have low risk of food insecurity.
   (b) Would you classify these countries as developing or developed?
4. **GS2** What does it mean to live in a country with low risk of food insecurity?
5. **GS2** The Food Security Index was based on evaluating five different indicators. Why do you think indicators such as accessibility and safety were included?
6. **GS1** What factors make people vulnerable to food insecurity?
7. **GS1** What is the difference between undernutrition and malnutrition?
8. **GS2** Explain how conflict can lead to food insecurity.

5.2 Exercise 2: Apply your understanding

1. **GS6** How do you think climate **change** might affect Australia's food security?
2. **GS6** What natural or human events could disrupt our food security?
3. **GS3** Refer to **FIGURE 3**.
   (a) With the use of dates and percentages, describe the main trend in food security throughout the world for 2014–2017.
   (b) Compare the trends in food security for Africa and North America/Europe over time. Use figures in your answer.
4. **GS6** Develop five steps you think would reduce a country's risk of food insecurity. Give reasons for your choices.
5. **GS6** How can Australia best help another country that is at high risk of having insufficient food for its people?
6. **GS6** At the turn of the twentieth century, the total worldwide spending on agricultural research was US$23 billion, compared with US$1.5 trillion on weapons. Do we have our priorities right? Write a short 'letter to the editor' outlining your view.

Try these questions in learnON for instant, corrective feedback. Go to www.jacplus.com.au.

5.3 SkillBuilder: Constructing and describing complex choropleth maps

What is a complex choropleth map?

A complex choropleth map is a map that is shaded or coloured to show the average density or concentration of a particular feature or variable, and it shows an area in detail.

Select your learnON format to access:

- an explanation of the skill (Tell me)
- a step-by-step process to develop the skill, with an example (Show me)
- an activity to allow you to practise the skill (Let me do it)
- questions to consolidate your understanding of the skill.

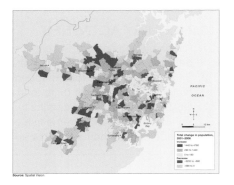

 **Resources**

🎞 **Video eLesson** Constructing and describing complex choropleth maps (eles-1732)

🧩 **Interactivity** Constructing and describing complex choropleth maps (int-3350)

5.4 Impacts of land loss on food security

5.4.1 How is land lost?

Land is absolutely essential for food production, and the world has more than enough arable land to meet future demands for food. Nevertheless, we need to find a balance between competing demands for this finite resource.

The loss of productive land has two main causes. First, there is the degradation of land quality through such things as erosion, **desertification** and salinity. Second, there is the competition for land from non-food crops, such as biofuels, and from expanding urban areas. As **FIGURE 1** shows, the growth in world population is inversely proportional to the amount of arable land available. This does not even take into consideration the land that is degraded and no longer suitable for growing food.

FIGURE 1 Comparison of world population growth and arable land per capita

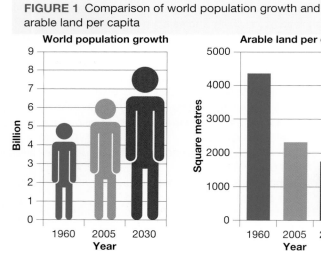

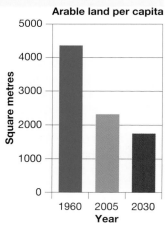

Land degradation

Although there have been significant improvements in crop yields, seeds, fertilisers and irrigation, they have come at a cost. Environmental degradation of water and land resources places future food production at risk.

The main forms of land degradation are:
- erosion by wind and water
- salinity
- pest invasion
- loss of biodiversity
- desertification.

Land degradation occurs in all food-producing biomes across the globe. Some degradation occurs naturally; for example, a heavy rainstorm can easily wash away topsoil. However, the most extensive degradation is caused by overcultivation, overgrazing, overwatering, overloading with chemicals and overclearing (see **FIGURE 2**). More than 75 per cent of the planet's land is considered degraded, impacting on the lives of more than 3 billion people. In China, erosion affects over 40 per cent of the land area; up to 10 million hectares are contaminated by pollutants.

FIGURE 2 Land degradation caused by deforestation in Madagascar

Competition for land

There has been a growing global trend to convert valuable cropland to other uses. Urban growth, industrialisation and energy production all require land. Melbourne currently produces enough food to supply 41 per cent its needs. With an estimated population of 7–8 million and the consequent growth in city size by the year 2050, the city will need 60 per cent more food. The capacity of current farmland will provide only 18 per cent of the city's needs.

Creeping cities

Cities tend to develop in places that are agriculturally productive. However, as they expand, they encroach on valuable farmland. Approximately 3 per cent of the world's land areas are urbanised, but this is expected to increase to 4–5 per cent by 2050.

FIGURE 3 Satellite image of New Delhi, India, in (a) 1989 and (b) 2018 — the expansion of the city has taken over valuable arable land.

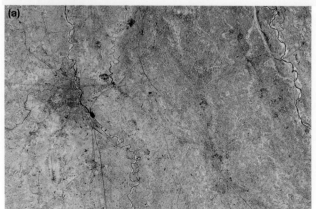
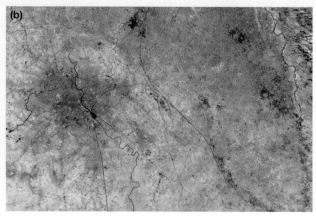

Growing fuel

Traditionally, the main forms of biofuel have been wood and charcoal. Almost 90 per cent of the wood harvested in Africa and 40 per cent of that harvested in Asia is used for heating and cooking. Today, people are seeking more renewable energy sources and they want to reduce CO_2 emissions associated with deforestation, so there is greater demand for alternative energy sources. Consequently, the use of agricultural crops to produce biofuels is increasing. Ethanol (mostly used as a substitute for petrol) is extracted from crops such as corn, sugar cane and cassava. Biodiesel is derived from plantation crops such as palm oil, soya beans and **jatropha**. The growth of the biofuel industry has the potential to threaten future food security by:

- changing food crops to fuel crops, so less food is produced and crops have to be grown on **marginal land** rather than arable land
- increasing prices, which makes staple foods too expensive for people to purchase
- forcing disadvantaged groups, such as women and the landless poor, to compete against the might of the biofuel industry.

Land grabs

A growing challenge to world food security is the purchase or lease of land, largely in developing nations, by resource-poor but wealthier nations. Large-scale 'land grabs', as they are known, have the potential to improve production and yields but at the same time there is growing concern over the loss of land rights and food security for local populations.

Since 2000, foreign investors have acquired over 26 million hectares around the world to produce food crops and biofuels. **FIGURE 4** shows the extent of China's expansion into other countries with investments in land and agricultural businesses.

FIGURE 4 Global map of China's land and food footprint

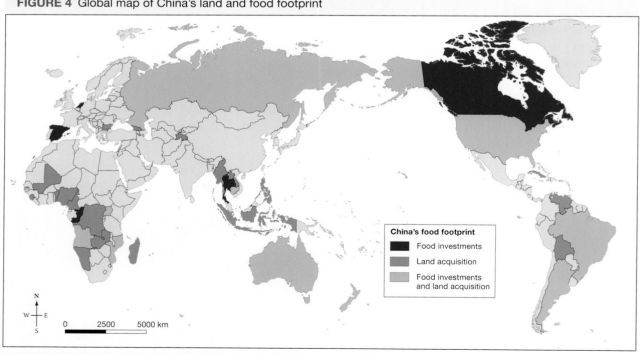

Source: The Heritage Foundation, GRAIN.org, Bloomberg. Map drawn by Spatial Vision.

Forty-two per cent of global acquisitions have occurred in Africa, examples of which can be seen in **FIGURE 5**. Africa's appeal is based on the fact that the continent accounts for 60 per cent of the world's arable land and yet most countries within it currently achieve less than 25 per cent of their potential yield.

FIGURE 5 Examples of land grabs in Africa

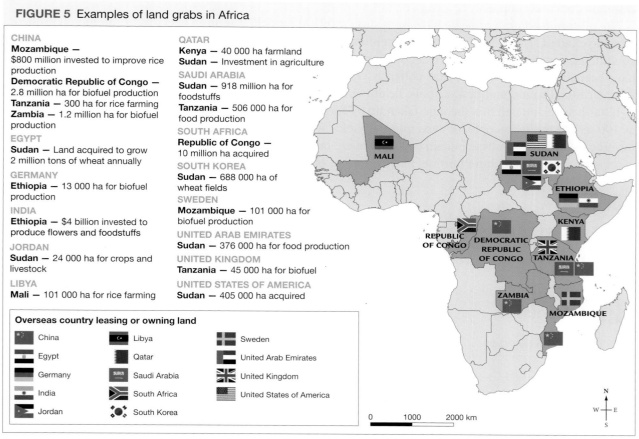

CHINA
Mozambique — $800 million invested to improve rice production
Democratic Republic of Congo — 2.8 million ha for biofuel production
Tanzania — 300 ha for rice farming
Zambia — 1.2 million ha for biofuel production

EGYPT
Sudan — Land acquired to grow 2 million tons of wheat annually

GERMANY
Ethiopia — 13 000 ha for biofuel production

INDIA
Ethiopia — $4 billion invested to produce flowers and foodstuffs

JORDAN
Sudan — 24 000 ha for crops and livestock

LIBYA
Mali — 101 000 ha for rice farming

QATAR
Kenya — 40 000 ha farmland
Sudan — Investment in agriculture

SAUDI ARABIA
Sudan — 918 million ha for foodstuffs
Tanzania — 506 000 ha for food production

SOUTH AFRICA
Republic of Congo — 10 million ha acquired

SOUTH KOREA
Sudan — 688 000 ha of wheat fields

SWEDEN
Mozambique — 101 000 ha for biofuel production

UNITED ARAB EMIRATES
Sudan — 376 000 ha for food production

UNITED KINGDOM
Tanzania — 45 000 ha for biofuel

UNITED STATES OF AMERICA
Sudan — 405 000 ha acquired

Overseas country leasing or owning land

China
Egypt
Germany
India
Jordan
Libya
Qatar
Saudi Arabia
South Africa
South Korea
Sweden
United Arab Emirates
United Kingdom
United States of America

MALI
SUDAN
ETHIOPIA
KENYA
REPUBLIC OF CONGO
DEMOCRATIC REPUBLIC OF CONGO
TANZANIA
ZAMBIA
MOZAMBIQUE

0   1000   2000 km

Source: Food and Agriculture Organization, International Food Policy Research Institute

The rise of land grabs came about as a result of the 'triple-F' crisis — food, fuel and finance.
- *Food crisis:* massive increases in world food prices in 2007–08 emphasised the need for those countries heavily reliant on importing food, such as Saudi Arabia and China, to improve their food security by obtaining land in other countries to produce food to meet their own needs.
- *Fuel crisis:* rising and fluctuating oil prices in 2007–09 created an incentive for countries to acquire land to produce their own biofuels (see **FIGURE 5**).
- *Financial crisis:* the global financial crisis in 2008 saw organisations switch from investing in stocks and shares to land in overseas countries, especially land that could be used to generate food and fuel crops.

The risk to food security

Investors in farmland are, understandably, seeking large expanses of land that has fertile soils and good rainfall or access to irrigation water. In many instances, land that is purchased is already occupied and used by small-scale farmers, often women who rarely benefit from any compensation. Prices for land can be much lower and there is frequently corruption, with much money going to local and government officials. People can also be forced off their land by governments keen to make deals with wealthy governments and corporations. Many land grabs have neglected the social, economic and environmental impacts of the deals.

With the purchase of land can come the right to withdraw the water linked to it and this can deny local people access to water for fishing, farming and watering animals. Withdrawal of water can reduce flow downstream. The Niger River, West Africa's largest river, flows through three countries and sustains over 100 million people, so any large-scale water reductions create significant impact to downstream environments and people.

Not all farmland grab projects have been successful. At least 17.5 million hectares of foreign-controlled land have failed. There are a number of interconnected reasons, including: a lack of understanding of local conditions; natural disasters; failed accounting; and, increasingly, challenges from local communities that have been displaced. When projects collapse, communities rarely get their lands back or are compensated for their loss. Promises of new schools, health clinics, infrastructure and jobs simply disappear.

It has been estimated that the land taken up by foreign investors for biofuel projects could feed as many as 190 to 370 million people, or even more, if yields were raised to the level of industrialised western farming. In addition to these human costs, there are important concerns about environmental risks that are associated with monoculture farming and the loss of biodiversity in the region.

DISCUSS
'Land grabs are the solution to establishing a country's food security.' Provide an argument for this viewpoint and an argument against this viewpoint. Ensure that your argument is supported with evidence and is logical.

[Critical and Creative Thinking Capability]

 Resources

 Google Earth New Delhi

5.4 EXERCISES

Geographical skills key: GS1 Remembering and understanding **GS2** Describing and explaining **GS3** Comparing and contrasting **GS4** Classifying, organising, constructing **GS5** Examining, analysing, interpreting **GS6** Evaluating, predicting, proposing

5.4 Exercise 1: Check your understanding
1. **GS1** What are the two main ways that productive farmland can be lost?
2. **GS1** Why is the use of corn as a biofuel a threat to food security?
3. **GS1** What is meant by the term *land grab*?
4. **GS2** Refer to **FIGURE 1**.
    (a) Describe the **changes** in population growth and the arable land per person between 1960 and 2030, making use of figures.
    (b) What do these graphs suggest about food security?
5. **GS3** Compare the advantages and disadvantages in developing and developed nations of using traditional biofuels, such as wood and charcoal, instead of oil and gas.

5.4 Exercise 2: Apply your understanding
1. **GS2** What is *jatropha*? What are the benefits of growing this rather than corn and other biofuels?
2. **GS6** Do you think Australia will need to purchase farmland overseas? Give reasons for your answer.
3. **GS6** Are land grabs the most effective solution for establishing a country's food security? Outline your view.
4. **GS6** Refer to **FIGURE 4**.
    (a) Describe the distribution of countries in which China has acquired land.
    (b) Suggest reasons why China might invest in food production and land in Australia.
5. **GS5** Refer to **FIGURE 5**. What do you notice about the use of land in Africa that is being acquired by foreign countries?

Try these questions in learnON for instant, corrective feedback. Go to www.jacplus.com.au.

5.5 SkillBuilder: Interpreting satellite images to show change over time

What is a satellite image?

A satellite image is an image taken from a satellite orbiting the Earth. Satellite images allow us to see very large areas — much larger than those that can be visualised using vertical aerial photography.

Select your learnON format to access:

- an explanation of the skill (Tell me)
- a step-by-step process to develop the skill, with an example (Show me)
- an activity to allow you to practise the skill (Let me do it)
- questions to consolidate your understanding of the skill.

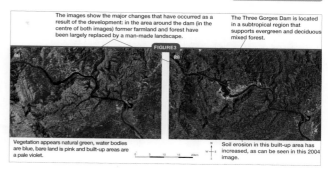

The images show the major changes that have occurred as a result of the development: in the area around the dam (in the centre of both images) former farmland and forest have been largely replaced by a man-made landscape.

The Three Gorges Dam is located in a subtropical region that supports evergreen and deciduous mixed forest.

FIGURE 3

Vegetation appears natural green, water bodies are blue, bare land is pink and built-up areas are a pale violet.

Soil erosion in this built-up area has increased, as can be seen in this 2004 image.

On Resources

Video eLesson Interpreting satellite images to show change over time (eles-1733)

Interactivity Interpreting satellite images to show change over time (int-3351)

5.6 Water — a vital part of the picture

5.6.1 Why are we running low on water?

There is no substitute for water. Without water there is no food, and agriculture already consumes 70 per cent of the world's fresh water. Every type of food production — cropping, grazing and processing — requires water. Thus, a lack of water is possibly the most limiting factor for increasing food production in future.

To feed an additional two billion people by 2050, the world will need to generate more food and use more water. The two main concerns that threaten future water security are water quantity and water quality.

In theory, the world has enough water; it is just not available where we want it or when we want it, and it is not easy to move from place to place. We already use the most accessible surface water, and now we are looking for it beneath our feet. Underground **aquifers** hold 100 times more water than surface rivers and lakes. However, groundwater is not always used at a sustainable rate, with extraction exceeding natural recharge, or filling. This occurs in many of the world's major food-producing places, in countries such as the United States, China and India.

Water insecurity is connected with food insecurity. **FIGURE 2** shows the predicted number of people who will face **water stress** and water scarcity in the future. A more complex view is seen in **FIGURE 3**, which shows an interconnection between increased demand for water and predicted climate change, population increase and greater industrialisation in the 2050s.

FIGURE 1 Water scarcity is a serious threat to food security.

FIGURE 2 People facing water stress and water scarcity

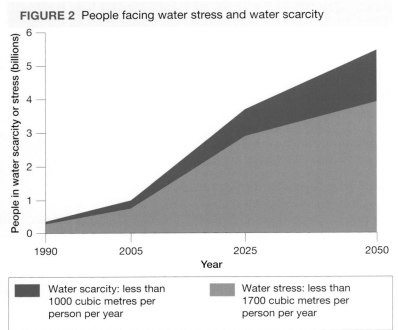

Water scarcity: less than 1000 cubic metres per person per year

Water stress: less than 1700 cubic metres per person per year

FIGURE 3 How water availability may change with temperature, population and industrialisation increase, 2050s

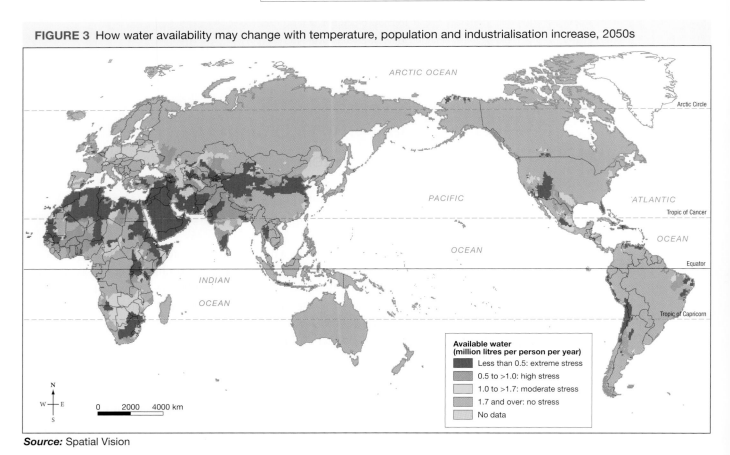

Available water (million litres per person per year)

- Less than 0.5: extreme stress
- 0.5 to >1.0: high stress
- 1.0 to >1.7: moderate stress
- 1.7 and over: no stress
- No data

Source: Spatial Vision

When water availability drops below 1.5 million litres per person per year, a country needs to start importing food; this makes the country vulnerable to changes in global prices. Developing countries that experience water stress cannot afford to import food. They are also more vulnerable to environmental disasters. In developing countries 70 per cent of food emergencies are brought on by drought.

The main causes of the growing water shortage are outlined below.
- *Food production.* It is estimated that an additional 6000 cubic kilometres of fresh water will be needed for irrigation to meet future food demand. Changes in diet, especially increased meat consumption, require more water to grow the crops and pasture that feed the animals. A typical meat eater's diet requires double the amount of water that a vegetarian diet requires.
- *Growth of urban and industrial demand.* Water for farming is diverted to urban populations, and productive land is converted to urban use.
- *Poor farming practices.* Water is wasted through inefficient irrigation methods and cultivating water-hungry crops such as rice. Poorly maintained irrigation infrastructure, such as pipes, canals and pumps, creates leakage.
- *Over-extraction.* Improved technology and cheaper, more available energy have enabled us to pump more groundwater from deeper aquifers. This is not always done at a sustainable rate, so as water is removed, less is available to refill lakes, rivers and wetlands.
- *Poor management.* Governments often price water cheaply, so irrigation schemes use water unsustainably. Some countries may have available water but lack the money to develop irrigation schemes.

5.6.2 Why is water quality deteriorating?

Agriculture is a major contributor to water pollution. Excess nutrients, pesticides, sediment and other pollutants can run off farmland or leach into soils, groundwater, streams and lakes. Excessive irrigation can cause waterlogging or soil salinity. This salty water not only poisons the soil but also drains into river systems. Industrial waste, untreated sewage and urban run-off also pollute water that may be used to irrigate farmland. Food that is irrigated with polluted water can actually pass on diseases and other medical problems, such as heavy-metal poisoning, to people. Pollution is an important contributor to the scarcity of clean, **potable** water.

Resources

🧩 **Interactivity** The last drop (int-3328)

🔗 **Weblinks** Water use

Water availability

5.6 EXERCISES

Geographical skills key: GS1 Remembering and understanding **GS2** Describing and explaining **GS3** Comparing and contrasting **GS4** Classifying, organising, constructing **GS5** Examining, analysing, interpreting **GS6** Evaluating, predicting, proposing

5.6 Exercise 1: Check your understanding

1. **GS2** Examine **FIGURE 2**.
   (a) Describe the projected *changes* in the number of people affected by water stress between 1990 and 2050. Use figures in your description.
   (b) How do these *changes* compare with figures for water scarcity?
2. **GS2** If a country has an average of 0.5 to <1.0 million litres of water per person per year, is it considered to be water stressed? Why?
3. **GS1** Why is agriculture both a contributor to and a victim of water pollution?
4. **GS2** Refer to **FIGURE 3**.
   (a) Describe those *places* in the world that are predicted to be in high to extreme water stress in the 2050s.
   (b) How could you explain why *places* like Eastern Europe could face water scarcity?
5. **GS2** Why would underground aquifers be able to hold more water than surface rivers and lakes?

5.7 Climate change challenges for food security

5.7.1 How will food security be affected by climate change?

The impacts of climate change on future world food security are a case of give and take. Some regions of the world will benefit from increases in temperature and rainfall, while others will face the threat of greater climatic uncertainty, lower rainfall and more frequent drought. In either case, food production will be affected.

Agriculture is important for food security, because it provides people with food to survive. It is also the main source of employment and income for 26 per cent of the world's workforce. In heavily populated countries in Asia, between 40 and 50 per cent of the workforce is engaged in food production, and this figure increases to an average of 54 per cent in sub-Saharan Africa.

It is difficult to predict the likely impacts of climate change, because there are many environmental and human factors involved (see **FIGURE 1**), as well as different predictions from scientists (see **FIGURE 2**).

FIGURE 1 Possible impacts of climate change on food production

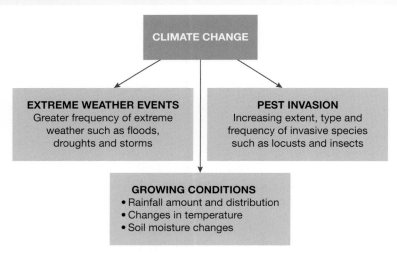

CLIMATE CHANGE

EXTREME WEATHER EVENTS
Greater frequency of extreme weather such as floods, droughts and storms

PEST INVASION
Increasing extent, type and frequency of invasive species such as locusts and insects

GROWING CONDITIONS
• Rainfall amount and distribution
• Changes in temperature
• Soil moisture changes

FIGURE 2 Projected consequences of climate change

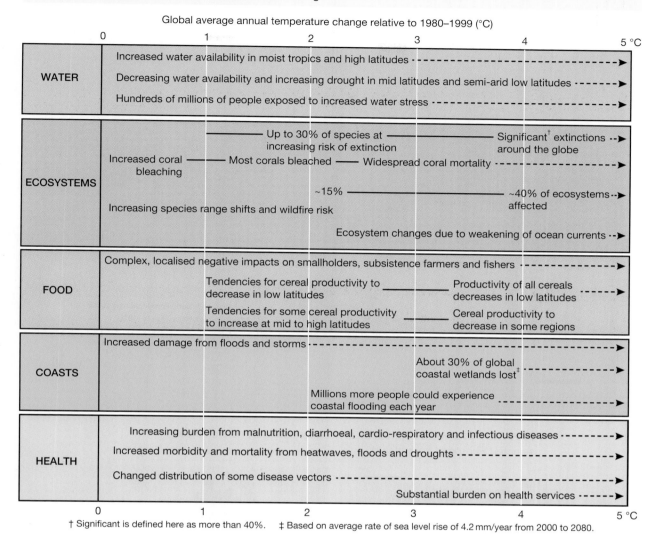

Global average annual temperature change relative to 1980–1999 (°C)

| | 0 | 1 | 2 | 3 | 4 | 5 °C |

WATER
- Increased water availability in moist tropics and high latitudes ----►
- Decreasing water availability and increasing drought in mid latitudes and semi-arid low latitudes ----►
- Hundreds of millions of people exposed to increased water stress ----►

ECOSYSTEMS
- Up to 30% of species at increasing risk of extinction ── Significant[†] extinctions ►around the globe
- Increased coral bleaching ── Most corals bleached ── Widespread coral mortality ----►
- ~15% ── ~40% of ecosystems ► affected
- Increasing species range shifts and wildfire risk
- Ecosystem changes due to weakening of ocean currents ►

FOOD
- Complex, localised negative impacts on smallholders, subsistence farmers and fishers ----►
- Tendencies for cereal productivity to decrease in low latitudes ── Productivity of all cereals decreases in low latitudes ----►
- Tendencies for some cereal productivity to increase at mid to high latitudes ── Cereal productivity to decrease in some regions

COASTS
- Increased damage from floods and storms ----►
- About 30% of global coastal wetlands lost[‡] ----►
- Millions more people could experience coastal flooding each year ----►

HEALTH
- Increasing burden from malnutrition, diarrhoeal, cardio-respiratory and infectious diseases ----►
- Increased morbidity and mortality from heatwaves, floods and droughts ----►
- Changed distribution of some disease vectors ----►
- Substantial burden on health services ----►

| | 0 | 1 | 2 | 3 | 4 | 5 °C |

† Significant is defined here as more than 40%. ‡ Based on average rate of sea level rise of 4.2 mm/year from 2000 to 2080.

There is a wide range of possible impacts of climate change. Sea-level rises may cause flooding and the loss of productive land in low-lying coastal areas, such as the Bangladesh and Nile River deltas. Changes in temperatures and rainfall may cause an increase in pests and plant diseases. However, agriculture is adaptable. Crops can be planted and harvested at different times, and new types of seeds and plants, or more tolerant species, can be used. Low-lying land may be lost, but higher elevations, such as mountain slopes, may become more suitable. The loss in productivity in some places may be balanced by increased production in other places. **FIGURE 3** demonstrates the effects of climate change on cereal crops, while **FIGURE 4** shows the range of potential impacts across Europe.

Essentially, hundreds of millions of people are at risk of increased food insecurity if they have to become more dependent on imported food. This will be evident in the poorer countries of Asia and sub-Saharan Africa, where agriculture dominates the economy. There is also a risk of greater numbers of **environmental refugees** or people fleeing places of food insecurity.

DISCUSS

Should food be shared more equitably around the world? How might this be achieved?   **[Ethical Capability]**

FIGURE 3 Predictions of the effects of climate change on cereal crops

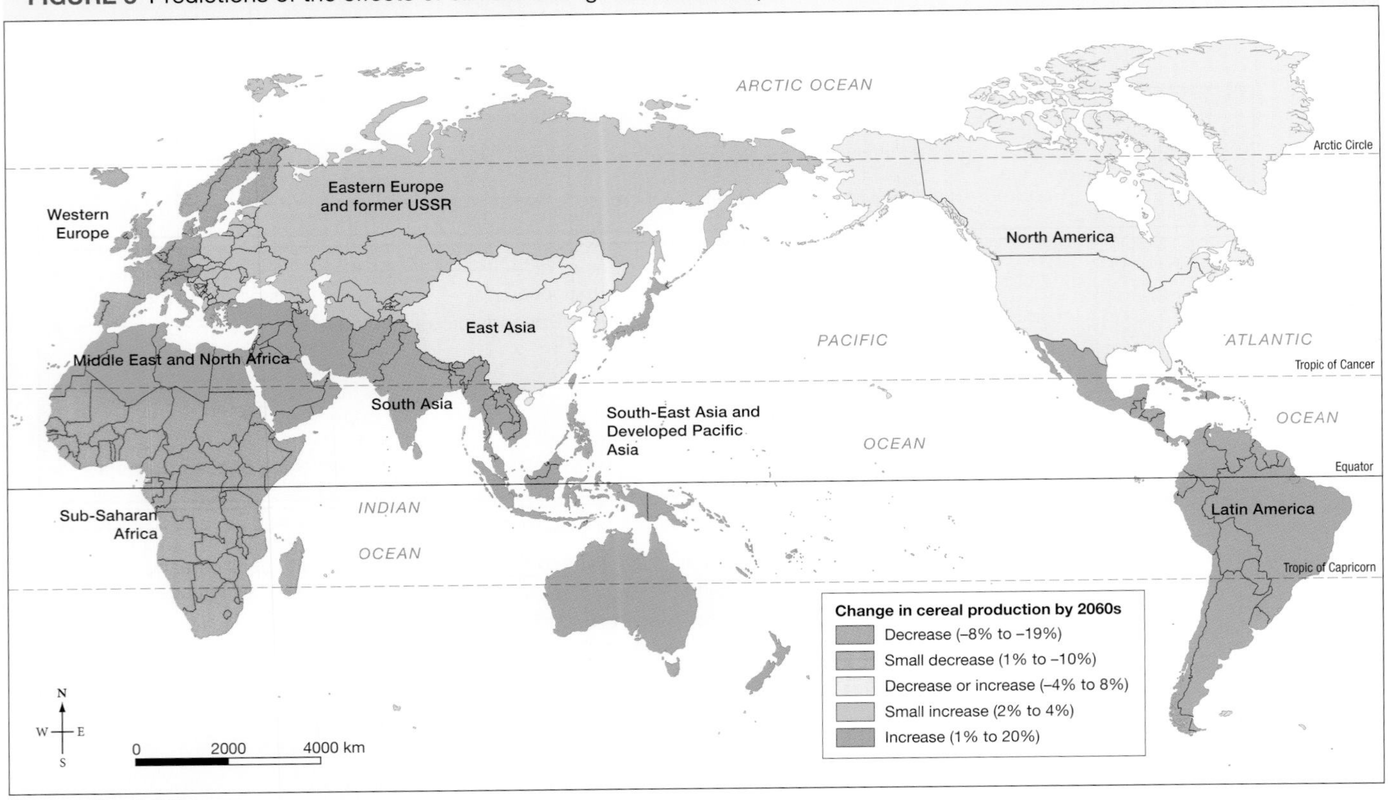

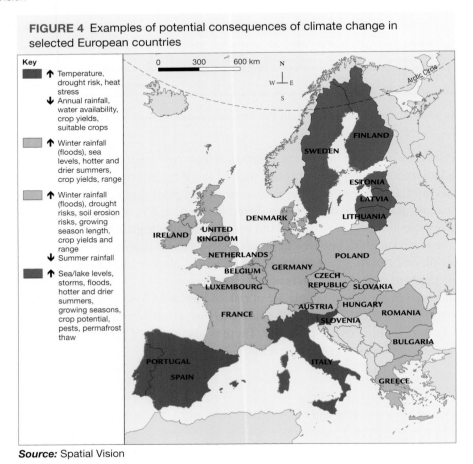

ARCTIC OCEAN

Arctic Circle

Eastern Europe and former USSR

Western Europe

North America

East Asia

PACIFIC

ATLANTIC

Middle East and North Africa

Tropic of Cancer

OCEAN

South Asia

South-East Asia and Developed Pacific Asia

OCEAN

Equator

INDIAN

Latin America

Sub-Saharan Africa

OCEAN

Tropic of Capricorn

Change in cereal production by 2060s
- Decrease (−8% to −19%)
- Small decrease (1% to −10%)
- Decrease or increase (−4% to 8%)
- Small increase (2% to 4%)
- Increase (1% to 20%)

N
W — E
S

0   2000   4000 km

Source: Spatial Vision

FIGURE 4 Examples of potential consequences of climate change in selected European countries

0   300   600 km

N
W — E
S

Arctic Circle

Key

↑ Temperature, drought risk, heat stress
↓ Annual rainfall, water availability, crop yields, suitable crops

↑ Winter rainfall (floods), sea levels, hotter and drier summers, crop yields, range

↑ Winter rainfall (floods), drought risks, soil erosion risks, growing season length, crop yields and range
↓ Summer rainfall

↑ Sea/lake levels, storms, floods, hotter and drier summers, growing seasons, crop potential, pests, permafrost thaw

FINLAND

SWEDEN

ESTONIA

LATVIA

LITHUANIA

DENMARK

IRELAND

UNITED KINGDOM

NETHERLANDS

POLAND

BELGIUM

GERMANY

CZECH REPUBLIC

SLOVAKIA

LUXEMBOURG

AUSTRIA

HUNGARY

ROMANIA

FRANCE

SLOVENIA

BULGARIA

PORTUGAL

SPAIN

ITALY

GREECE

Source: Spatial Vision

5.7 INQUIRY ACTIVITIES

1. Research potential impacts of climate **change** on Australia. Create an annotated map to illustrate your findings.
   Classifying, organising, constructing
2. Use the **How to feed the world in 2050** weblink in the Resources tab to find out more about the impact of climate **change** on food security.
   Examining, analysing, interpreting

5.7 EXERCISES

Geographical skills key: GS1 Remembering and understanding **GS2** Describing and explaining **GS3** Comparing and contrasting **GS4** Classifying, organising, constructing **GS5** Examining, analysing, interpreting **GS6** Evaluating, predicting, proposing

5.7 Exercise 1: Check your understanding

1. **GS2** Describe the **interconnection** between **environmental** refugees and climate **change**.
2. **GS2** Explain how the impacts of climate **change** may be a benefit to food production in a particular **place**.
3. **GS2** How would an increase in extreme weather events impact on food production?
4. **GS2** How might technologies such as glasshouses and irrigation help reduce the impacts of global warming?
5. **GS2** Why are the impacts of climate **change** likely to be felt more in those countries with a high percentage of their population in the agricultural workforce?

5.7 Exercise 2: Apply your understanding

1. **GS6** How might a country such as Australia best prepare its food production systems to cope with potential **changes** in climate?
2. **GS5** Refer to **FIGURE 2** and decide whether the following statements are true or false.
   (a) If temperatures increase by 3 °C, crop yields around the equator would rise.
   (b) **Changes** in extreme weather events are unlikely unless temperatures increase by at least 1 °C.
   (c) Food insecurity will be felt greatly in developing regions if temperatures rise more than 4 °C.
   (d) **Places** that are likely to experience decreasing crop yields will be found in the higher latitudes.
3. **GS5** Refer to **FIGURE 3**.
   (a) Which **places** have the potential to be grain exporters and which **places** are likely to become dependent on grain imports? Use data in your answer.
   (b) What are the economic and social implications of this for countries in these regions?
4. **GS5** Refer to **FIGURE 4**.
   (a) Which countries of Europe will benefit from climate **change** in terms of food production and which countries are likely to suffer negative outcomes?
   (b) Would increased irrigation be a **sustainable** solution to growing food in Spain? Explain your answer.
5. **GS4** Refer to **FIGURE 3** to complete the table below, classifying each of the following countries according to their predicted **change** to cereal production: Bangladesh, Brazil, England, Germany, Indonesia, Mexico, South Africa, South Korea.

Decrease (−8% to −19%)	Small decrease (1% to −10%)	Decrease or increase (−4% to 8%)	Small increase (2% to 4%)	Increase (1% to 20%)

Try these questions in learnON for instant, corrective feedback. Go to www.jacplus.com.au.

5.8 Managing food wastage

5.8.1 What is the link between waste and food security?

What food have you thrown out today? Across the world, one-third of all food produced is wasted. Each year, around 1.6 million tonnes of food, worth up to $1.2 trillion, is dumped while more than 850 million people remain undernourished. According to the United Nations' Food and Agriculture Organization, one-quarter of the food wasted each year could feed all of the world's hungry people.

To meet the growing demand for food by the middle of this century, it has been calculated that the world will need to produce as much food as has been produced over the past 8000 years. Although the world does produce sufficient food for everyone, distribution and affordability prevent it from getting to everyone who needs it. However, dealing with food wastage could certainly help to reduce food vulnerability.

FIGURE 1 Surplus tomatoes dumped in Tenerife, Canary Islands

Food wastage also represents a waste of the resources used in production, such as land, fertiliser and energy. Waste can increase prices, making food less affordable. The World Bank has calculated that in sub-Saharan Africa, a region prone to food insecurity, a reduction of only one per cent wastage could save $40 million per year, with most of this saving going to the farmers.

A consequence of food wastage is the need to dispose of the waste, usually by dumping or burning. Food waste now contributes 8 per cent of global greenhouse gas emissions.

5.8.2 Where and why is food being wasted?

Food waste exists in all countries, regardless of their levels of development, although the causes of wastage vary. **FIGURE 2** shows the breakdown of food wastage on a regional basis.

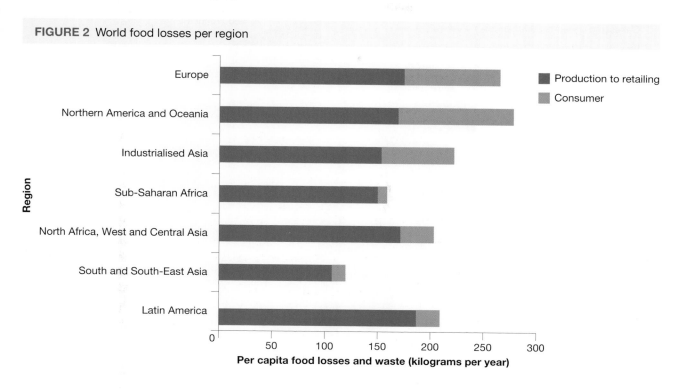

FIGURE 2 World food losses per region

In developing nations, food losses are mostly related to a lack of food-chain infrastructure and a poor knowledge of, or investment in, storage technologies on farms. Other causes of waste are: lack of refrigeration; limited or non-existent road and rail networks to deliver food to markets; and a shortage of processing and packaging facilities. In India, up to 40 per cent of fresh food is lost due to a lack of cold storage in wholesale and retail outlets. Over one-third of the rice harvest in South-East Asia can be destroyed by pests or spoilage.

In contrast, in the developed world, food waste is more evident at the retail and home stages of the food chain. In this case, food is relatively cheap so there is little incentive to avoid waste. Consumers are used to purchasing food that is visually appealing and unblemished, so retailers end up throwing out perfectly edible, if slightly damaged, food. More and more people rely on 'use by' dates, so despite the food still being suitable to eat, it is discarded. Waste is also a part of the growing culture of 'supersize' or 'buy one get one free' advertising. Further waste can occur if the discarded food is sent to landfill when it could be used for animal feed or even compost.

What is wasted in Australia?

Australia produces enough food for 60 million people, and this enables us to trade the surplus. Yet each person wastes an average of 361 kg of food each year. This costs the economy $20 billion annually. At the same time, four million Australians have experienced some form of food insecurity in the past year. This means that around 18 per cent of the population have not had enough food for themselves and their family, or could not afford to purchase food at some stage over the twelve-month period.

Within Victoria, food wastage costs $5.4 billion annually. The average household throws away $42 worth of food per week. **FIGURE 3** shows the composition of the 255 000 tonnes of food thrown into rubbish bins in Victoria each year.

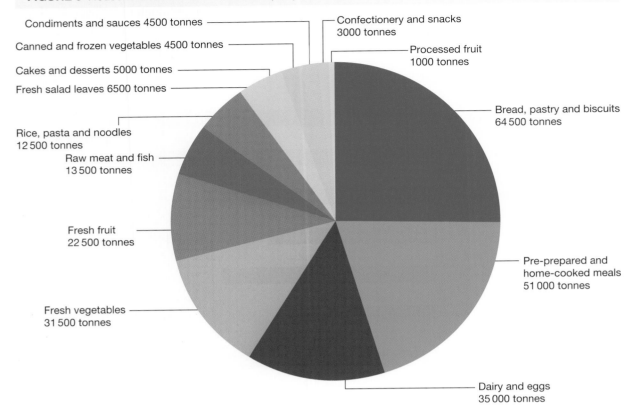

FIGURE 3 Household food waste in Victoria, per year

Condiments and sauces 4500 tonnes

Canned and frozen vegetables 4500 tonnes

Cakes and desserts 5000 tonnes

Fresh salad leaves 6500 tonnes

Rice, pasta and noodles 12500 tonnes

Raw meat and fish 13500 tonnes

Fresh fruit 22500 tonnes

Fresh vegetables 31500 tonnes

Confectionery and snacks 3000 tonnes

Processed fruit 1000 tonnes

Bread, pastry and biscuits 64500 tonnes

Pre-prepared and home-cooked meals 51000 tonnes

Dairy and eggs 35000 tonnes

5.8.3 What can be done about food waste?

Reducing global food waste is a part of the new Sustainable Development Goals, a set of targets designed to develop a more sustainable future for the world. The specific target is to cut per capita food waste by 50 per cent by 2030. If this can be achieved, food security will be improved, greenhouse gases can be reduced, and valuable land and water resources will not be wasted.

Here is a snapshot of what is happening around the world:

- Farmers in Ghana are trialling a new phone app that shows farmers, food transporters and traders the fastest route to market, which reduces food spoilage. In addition, the app can identify illegal roadblocks set up to take bribes from drivers.
- In France, an estimated 10 million tonnes of food is wasted each year. A new law now compels restaurants to provide containers in which customers can take home uneaten food. Shops are also banned from destroying food products, and supermarkets must give away unsold food that has reached its use-by date, for distribution to charities. By 2020, all Parisian households should have a biowaste recycling bin for food scraps. Waste will be collected and converted into fertiliser or biofuels.
- Seoul in South Korea has taken a different approach in an effort to reduce its food waste by 20 per cent. It is trialling a program whereby people are charged according to the weight of the garbage they produce. The more kilograms generated, the higher the bill. In South Korea 95 per cent of food waste is recycled into compost, animal feed or fuel. Landfilling of food waste is banned.
- Australia has now set a target to reduce the amount of food waste by 50 per cent by 2030. Much of this will come from supporting food rescue operations such as Second Bite and Foodbank Australia (see subtopic 14.7). These organisations collect and redistribute surplus food. Foodbank provides relief to 710000 Australians every month, 26 per cent of whom are under 19 years old.

5.8 INQUIRY ACTIVITIES

1. In groups, and wearing disposable gloves, conduct a survey of the school rubbish bins after lunch. You may need to lay out newspaper onto which you can tip the contents of the bins. Some groups could also deal with food litter around the grounds.
   (a) Construct a table so that you can record the different food types, such as fruit, cakes, biscuits and so on.
   (b) Collate your results with the other groups in your class, and then graph your data.
   (c) Write a summary of your findings. What food types were most and least represented and why?
   (d) If your school has a canteen, ask the manager to address the class and talk about issues such as wastage, use-by dates and health department regulations.
   (e) You could also do a home bin audit and follow the same procedure.   **Examining, analysing, interpreting**
2. Visit a local food store, such as a supermarket, fresh food market, greengrocer or butcher. Interview a staff member and find out what happens to their food waste. Report back to the class.
   Examining, analysing, interpreting
3. Design a poster or short animation to inform other school members about the issue of food waste.
   Classifying, organising, constructing

5.8 EXERCISES

Geographical skills key: GS1 Remembering and understanding **GS2** Describing and explaining **GS3** Comparing and contrasting **GS4** Classifying, organising, constructing **GS5** Examining, analysing, interpreting **GS6** Evaluating, predicting, proposing

5.8 Exercise 1: Check your understanding

1. **GS2** Why is food waste a global problem?
2. **GS2** Explain the *interconnection* between food waste and global warming.
3. **GS2** Suggest ways in which food can be wasted or spoiled between production and retailing (between the farm gate and the supermarket).
4. **GS2** Why is there more food wasted by retailers and in homes in developed countries than in developing countries?
5. **GS6** In an effort to reduce food wastage, Woolworths and Coles supermarkets have established, respectively, 'The Odd Bunch' and 'I'm perfect' programs, through which bags of misshapen fruit and vegetables are sold at a reduced price. How effective do you think these programs might be? Explain your view.

5.8 Exercise 2: Apply your understanding

1. **GS6** Write a letter to the local newspaper voicing your thoughts on food wastage.
2. **GS5** Refer to **FIGURE 2**. Which regions of the world are shown to waste the greatest amount of food in the production-to-retailing and consumer sectors? Use data in your answer.
3. **GS5** Refer to **FIGURE 3**.
   (a) What are the three largest categories of food wasted?
   (b) Estimate the total amount that is wasted in these three groups.
   (c) Suggest reasons why these are the largest groupings.
4. **GS3** Consider South Korea's and Australia's plans to reduce food waste.
   (a) In table form, use a dot point summary to compare the strengths and weaknesses of the plans.
   (b) Which of the two plans do you think will be most effective, and why?
5. **GS6** Goal 12 of the United Nations Sustainable Development Goals aims to 'by 2030, halve per capita global food waste at the retail and consumer levels and reduce food losses along production and supply chains ...'. Do you think this is possible? Why or why not?

Try these questions in learnON for instant, corrective feedback. Go to www.jacplus.com.au.

5.9 Thinking Big research project: Famine crisis report

online only

SCENARIO
As a world-leading specialist in food security, the United Nations (UN) has asked you to write a report that will assist them in organising a response to a famine. You will need to present your report to a famine taskforce panel at the UN headquarters in New York.

Select your learnON format to access:
- the full project scenario
- details of the project task
- resources to guide your project work
- an assessment rubric.

 Resources

💡 **ProjectsPLUS** Thinking Big research project: Famine crisis report (pro-0191)

5.10 Review

online only

5.10.1 Key knowledge summary
Use this dot point summary to review the content covered in this topic.

5.10.2 Reflection
Reflect on your learning using the activities and resources provided.

 Resources

 eWorkbook Reflection (doc-31720)

Crossword (doc-31721)

 Interactivity Challenges to food security crossword (int-7647)

KEY TERMS
aquifer a body of permeable rock below the Earth's surface, which contains water, known as groundwater

arable describes land that can be used for growing crops

desertification the transformation of arable land into desert, which can result from climate change or from human practices such as deforestation and overgrazing

environmental refugees people who are forced to flee their home region due to environmental changes (such as drought, desertification, sea-level rise or monsoons) that affect their wellbeing or livelihood

indicator something that provides a pointer, especially to a trend

jatropha any plant of the genus *Jatropha*, but especially *Jatropha curcas*, which is used as a biofuel

malnourished describes someone who is not getting the right amount of the vitamins, minerals and other nutrients to maintain healthy tissues and organ function

marginal land describes agricultural land that is on the margin of cultivated zones and is at the lower limits of being arable

potable drinkable; safe to drink

undernourished describes someone who is not getting enough calories in their diet; that is, not enough to eat

water stress situation that occurs when water demand exceeds the amount available or when poor quality restricts its use

6 Meeting our future global food needs

6.1 Overview

Will there come a time when we don't have enough food to feed everyone?

6.1.1 Introduction

Currently we produce enough food to adequately feed everyone in the world. However, it is estimated that approximately one in every nine people (around 820 million) are going hungry. The world's population is expected to grow by another two billion people in the next 30 years. If we want to stop the number of hungry people from increasing, we will need improvements in food production, new sources of food, better aid programs, and different attitudes to food consumption and waste.

 Resources

   ☑ **eWorkbook**   Customisable worksheets for this topic

   ▤ **Video eLesson** Future food (eles-1721)

To access a pre-test and starter questions and receive immediate, **corrective feedback** and **sample responses** to every question, select your learnON format at www.jacplus.com.au.

6.2 Can we feed the future world population?

6.2.1 The prevalence and impacts of hunger

According to the World Health Organization, over 1.9 billion adults in the world are overweight, while 821 million go hungry each day. What can we do to change this imbalance and ensure equal, sustainable access to food for people across the globe?

The impact of hunger on people cannot be overstated. Hunger kills more people each year than malaria, AIDS and tuberculosis combined. It is estimated that we will need to produce between 70 and 100 per cent more food in order to feed future populations. New ideas, knowledge and techniques will be needed if we do not want millions more people to suffer malnourishment, starvation and vulnerability to disease. The challenge, though, is to do this in a way that is also sustainable. Population growth and limited supplies of arable land will affect how much food can be produced.

6.2.2 Challenges to food production

The distribution of the world's population and the availability of arable land per person is uneven. Regions with the fastest-growing future populations (see **FIGURE 1**) are also those where there is limited arable land per person.

FIGURE 1 Global population growth and percentage of total population growth by region, 2010–50

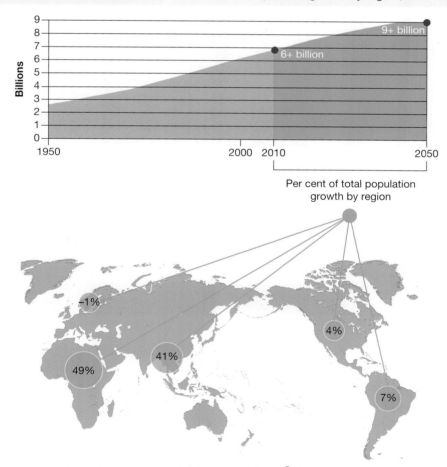

Source: Redrawn from an image by Global Harvest Initiative (*2011 GAP Report®: Measuring Global Agricultural Productivity*), data from the United Nations

One solution to feeding people who live in crowded spaces, such as Asia, or in environmentally challenging spaces, such as sub-Saharan Africa, is to increase the amount of trade in food products. This will involve moving food from places with crop surpluses (North America, Australia and Europe) to regions that are crowded or less productive. This means there will be an increase in the interconnection between some countries.

Preventing hunger on a global scale is important, but action also needs to be taken on a local scale. Over 70 per cent of the world's poor live in rural areas; improving their lives would create greater food security. If poor farmers can produce more food, they can feed themselves and provide for local markets. Improved infrastructure, such as roads in rural regions, would enable them to transport their produce to market and increase their incomes.

Factors affecting food production

Farming is a complex activity, and farmers around the world face many challenges in producing enough food to feed themselves and to create surpluses they can sell to increase their incomes. Some of these are outlined in **FIGURE 2**.

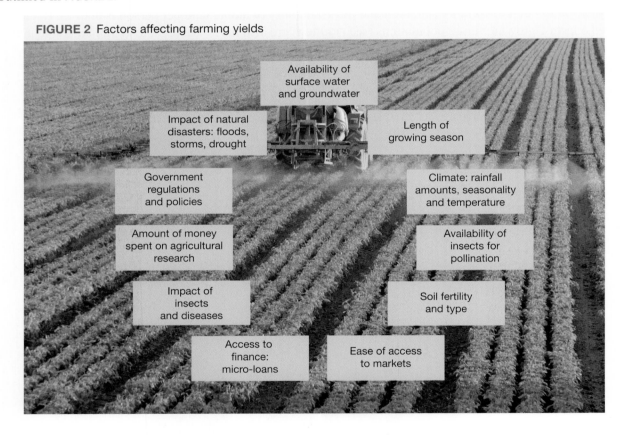

FIGURE 2 Factors affecting farming yields

Availability of surface water and groundwater

Impact of natural disasters: floods, storms, drought

Length of growing season

Government regulations and policies

Climate: rainfall amounts, seasonality and temperature

Amount of money spent on agricultural research

Availability of insects for pollination

Impact of insects and diseases

Soil fertility and type

Access to finance: micro-loans

Ease of access to markets

As urban areas grow, the amount of available arable land decreases. According to the United Nations Food and Agriculture Organization (FAO), the world has an extra 2.8 billion hectares of unused potential farmland. This is almost twice as much as is currently farmed. However, only a fraction of this extra land is realistically available for agricultural expansion, owing to inaccessibility and the need to preserve forest cover and land for infrastructure.

As mentioned, the growing populations of the future will be found in places where expansion of land for agriculture is already limited. Consequently, increased food production will need to come from better use of current agricultural areas, better use of technology, and new ways of thinking about food production and approaches to farming. One such example is the Ord River irrigation scheme in the East Kimberley region of Western Australia, which is transforming this semi-arid region and providing food in huge quantities for our Asian neighbours.

FIGURE 3 The Ord River Irrigation Scheme has alllowed great expansion of the available farming area in the region.

6.2 INQUIRY ACTIVITY

As well as affecting people's health, a shortage of food can have social and political effects. Undertake research into the series of food riots that occurred in a number of countries around the world in 2015.
a. Where did these riots occur?
b. What were the causes of these riots?
c. Why might governments need to prevent this situation from occurring again?

Examining, analysing, interpreting

6.2 EXERCISES

Geographical skills key: GS1 Remembering and understanding **GS2** Describing and explaining **GS3** Comparing and contrasting **GS4** Classifying, organising, constructing **GS5** Examining, analysing, interpreting **GS6** Evaluating, predicting, proposing

6.2 Exercise 1: Check your understanding

1. **GS1** How much more food is it estimated that we will need to produce in order to feed future populations?
2. **GS2** Explain why hunger is such a serious issue.
3. **GS1** What is the relationship between areas with fast-growing populations and the amount of arable land per person?
4. **GS1** What proportion of the world's poor live in rural areas?
5. **GS1** What may need to happen to ensure there is enough food in the future for people who live in *places* with growing populations and limited arable land?

6.2 Exercise 2: Apply your understanding

1. **GS5** Examine **FIGURE 1**.
   (a) Which region is predicted to decrease in population by 2050?
   (b) Which two continents are expected to have the greatest increase in population?
   (c) What is the predicted world population in 2050?
2. **GS6** Lack of food has caused people to leave their homes and move to cities in search of employment and food. Predict the *places* of the world where this is most likely to happen.
3. **GS2** How does a growing world population put pressure on food supplies?
4. **GS4** Classify the factors affecting farming yields as either *environmental,* economic, or social/political.
5. **GS6** Choose four of the factors affecting farming yields and suggest how these factors impact production levels.

Try these questions in learnON for instant, corrective feedback. Go to www.jacplus.com.au.

6.3 Improving food production

6.3.1 Finding ways of producing more

There are many strategies that can be used to create greater efficiencies and increased food production. **FIGURE 1** summarises some of these.

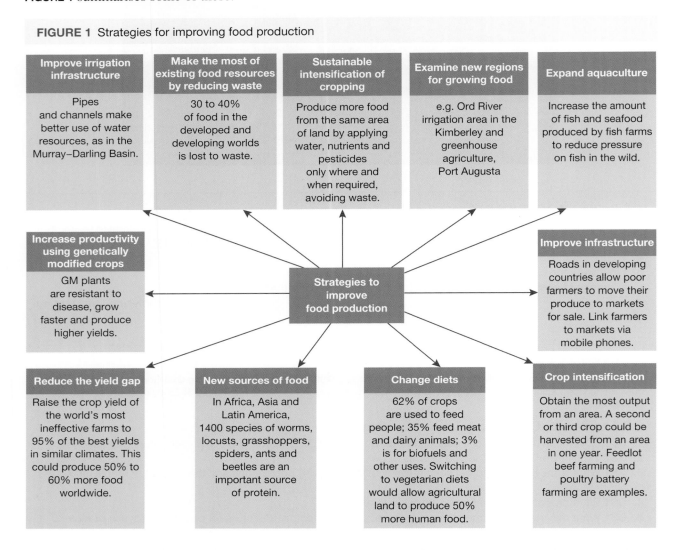

FIGURE 1 Strategies for improving food production

Improve irrigation infrastructure

Pipes and channels make better use of water resources, as in the Murray–Darling Basin.

Make the most of existing food resources by reducing waste

30 to 40% of food in the developed and developing worlds is lost to waste.

Sustainable intensification of cropping

Produce more food from the same area of land by applying water, nutrients and pesticides only where and when required, avoiding waste.

Examine new regions for growing food

e.g. Ord River irrigation area in the Kimberley and greenhouse agriculture, Port Augusta

Expand aquaculture

Increase the amount of fish and seafood produced by fish farms to reduce pressure on fish in the wild.

Increase productivity using genetically modified crops

GM plants are resistant to disease, grow faster and produce higher yields.

Strategies to improve food production

Improve infrastructure

Roads in developing countries allow poor farmers to move their produce to markets for sale. Link farmers to markets via mobile phones.

Reduce the yield gap

Raise the crop yield of the world's most ineffective farms to 95% of the best yields in similar climates. This could produce 50% to 60% more food worldwide.

New sources of food

In Africa, Asia and Latin America, 1400 species of worms, locusts, grasshoppers, spiders, ants and beetles are an important source of protein.

Change diets

62% of crops are used to feed people; 35% feed meat and dairy animals; 3% is for biofuels and other uses. Switching to vegetarian diets would allow agricultural land to produce 50% more human food.

Crop intensification

Obtain the most output from an area. A second or third crop could be harvested from an area in one year. Feedlot beef farming and poultry battery farming are examples.

The strategy that is likely to be the most important in increasing future crop production is the reduction of the **yield gap**. This means that farmers who are currently less productive will need to implement farming methods that will lead to increased yields so that their outputs are closer to those of more productive farmers. There is a serious yield gap in more than 157 countries (see **FIGURE 2**). If this gap could be closed, larger amounts of food would be available without the need for more land. There are wide geographic variations in crop and livestock productivity. Brazil, Indonesia, China and India have all made great progress in increasing their agricultural output. Much of the increase has been achieved through more efficient use of water and fertilisers.

The use of **genetically modified** (GM) foods has increased, and this has also increased crop yields. However, there is some opposition to GM crops because of concerns about:

- loss of seed varieties
- potential risks to the environment and people's health
- the fact that large companies hold the copyright to the seeds of GM plants that are food sources.

FIGURE 2 Yield gap for a combination of major crops, 2015

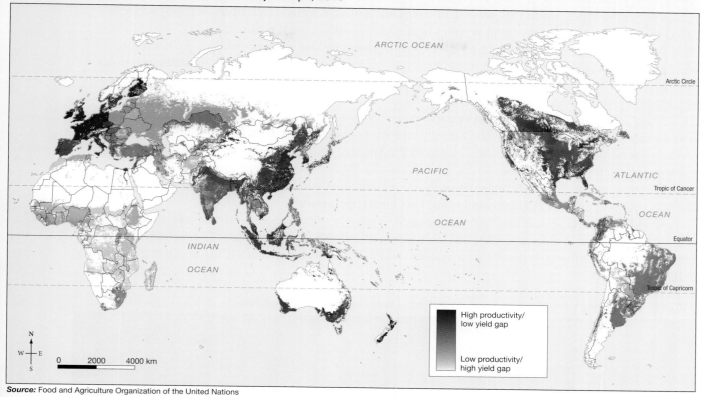

High productivity/ low yield gap

Low productivity/ high yield gap

Source: Food and Agriculture Organization of the United Nations

FIGURE 3 Global area of genetically modified crops in industrialised and developing countries 1996–2017

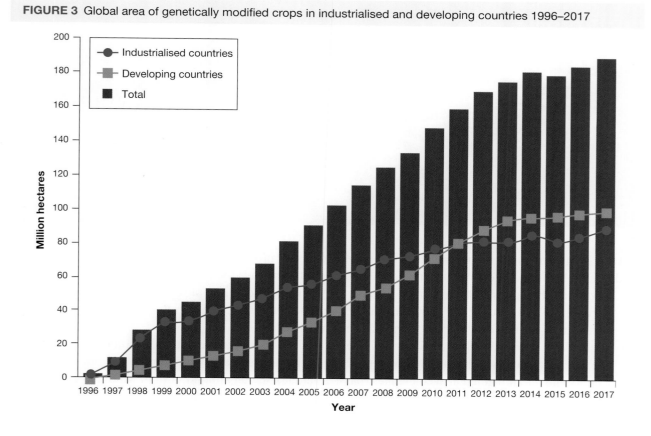

6.3.2 Innovative solutions

Because agriculture uses around 70 per cent of the planet's increasingly scarce freshwater resources, any method that can produce food without needing fresh water at all is a great advance.

Port Augusta is located in a hot, arid region of South Australia, and is not normally associated with agriculture. However, one company, Sundrop Farms, is using this region's abundant renewable resources of sunlight and sea water to produce high-quality, pesticide-free vegetables, including tomatoes, capsicums and cucumbers, and it does so all year round.

In 2016, a 20-hectare greenhouse was opened, powered by a 115 m solar tower with 23 000 mirrors. The mirrors concentrate the sun's energy and the collected heat creates steam to drive electricity production, heat the greenhouse, and desalinate sea water from the Spencer Gulf, producing up to one million litres of fresh water a day for crop irrigation.

The greenhouse aims to satisfy approximately 10 per cent of Australia's truss tomato demand and its sustainably farmed produce is already being sold at Coles supermarkets.

It is hoped that this type of technology can be used in many more places in Australia and around the world that have hot, arid climates previously considered unsuitable for horticulture. The technology has the potential to supply millions of people with healthy food in a sustainable manner while also using minimal fossil fuel resources.

FIGURE 4 The world's first Sundrop Farm is situated in Port Augusta, South Australia.

Australian farmers see technology as a means of decreasing production costs and increasing crop production. Additional technologies in Australian agriculture include the following.

- Robots are being tested to determine whether they can be used in complex jobs such as watering or harvesting. This would be of advantage in the horticultural sector, which is the third largest sector in agriculture, with an export trade worth $2.2 billion in 2017–18.
- Technology such as satellite positioning is being used to determine the optimal amounts of fertiliser to use on crop farms, which could increase profitability by as much as 14 per cent.
- Robots and an unmanned air vehicle have passed field tests at an almond farm in Mildura, Victoria. They are fitted with vision, laser, radar and conductivity sensors — including GPS and thermal sensors.

Resources

Interactivity More, or less, food (int-3329)

Weblink Vertical farming

6.3 INQUIRY ACTIVITY

Use the **Vertical farming** weblink in the Resources tab to watch a video clip on this topic.
a. What is being suggested about *environmentally sustainable* farming in the future?
b. Draw a diagram to show what a future vertical farm might look like.
c. How might vertical farms help to feed future populations?   **Examining, analysing, interpreting**

6.3 EXERCISES

Geographical skills key: GS1 Remembering and understanding **GS2** Describing and explaining **GS3** Comparing and contrasting **GS4** Classifying, organising, constructing **GS5** Examining, analysing, interpreting **GS6** Evaluating, predicting, proposing

6.3 Exercise 1: Check your understanding

1. **GS1** What is meant by the term *yield gap* and why is it important that this gap be narrowed?
2. **GS1** List three different strategies, other than closing the yield gap, for improving food production.
3. **GS1** What is meant by the term *genetically modified* (GM)?
4. **GS2** Explain the advantages and disadvantages of locating the large greenhouse near Port Augusta.
5. **GS2** Select one of the strategies outlined in **FIGURE 1** that can be used to improve food production. Explain this strategy in your own words and outline some of the strengths and weaknesses of this strategy.

6.3 Exercise 2: Apply your understanding

1. **GS5** Study **FIGURE 2**.
   (a) Which *places* have the highest yield gaps?
   (b) Which *places* have the lowest yield gaps?
2. **GS5** Examine **FIGURE 3**. What *changes* have there been in the production of genetically modified foods in (a) industrialised countries and (b) developing countries? Use the data from the graph to support your answer.
3. **GS6** Predict what the impact might be on people and *places* if the greenhouse method of farming shown in **FIGURE 4** were to become more readily available. What might be the effects on *places* where the yield gap is large compared to *places* that are currently more productive?
4. **GS4** Annotate key aspects of the Port Augusta landscape (greenhouses, solar collector, sandy soil, flat and barren landscape) shown in the photograph in **FIGURE 4**.
5. **GS6** Many Australian cities have large housing estates on their outskirts. This land was often used for market gardens or farmland. What impact might the loss of this productive land have on the price of food?
6. **GS6** Some areas of Australia that are currently national parks or marine parks may be sought after as agricultural land in the future. Outline your views on this.

Try these questions in learnON for instant, corrective feedback. Go to www.jacplus.com.au.

6.4 Global food aid

6.4.1 Who needs food aid and how is it delivered?

Food aid is food, money, goods and services given by wealthier, more developed nations to less developed nations for the specific purpose of helping those in need.

People who need food aid include:

- poor people who cannot buy food even if it is available, as they are often trapped in a cycle of hunger and poverty
- people who have fled violence or civil conflict
- people devastated by natural disasters.

There are three general categories of food aid:

1. Program food aid, which is organised between national governments and provides resources that offer budgetary support to countries in need
2. Project food aid, which is targeted at specific areas or groups and provides support for disaster prevention activities and poverty alleviation measures
3. Relief (crisis or emergency) food aid, which assists victims of man-made and natural disasters

The United Nations World Food Programme (WFP) is a voluntary arm of the United Nations. It reaches more than 80 million people, in more than 92 countries, with food assistance after disasters and conflicts. The WFP provides different types of food aid to people after natural disasters such as cyclones, floods and earthquakes. Some relief aid is provided in the short term as emergency food. Project food relief is often required over lengthy periods, typically after civil war or prolonged drought.

FIGURE 1 How the WFP works

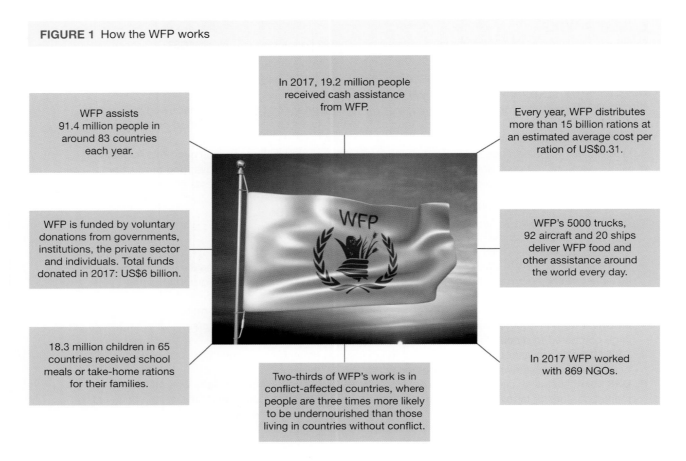

In 2017, 19.2 million people received cash assistance from WFP.

WFP assists 91.4 million people in around 83 countries each year.

Every year, WFP distributes more than 15 billion rations at an estimated average cost per ration of US$0.31.

WFP is funded by voluntary donations from governments, institutions, the private sector and individuals. Total funds donated in 2017: US$6 billion.

WFP's 5000 trucks, 92 aircraft and 20 ships deliver WFP food and other assistance around the world every day.

18.3 million children in 65 countries received school meals or take-home rations for their families.

Two-thirds of WFP's work is in conflict-affected countries, where people are three times more likely to be undernourished than those living in countries without conflict.

In 2017 WFP worked with 869 NGOs.

Natural and **anthropogenic** disasters are the drivers of hunger and malnutrition. The WFP works to prevent, mitigate and prepare for such disasters. In 2018, the WFP worked on five major humanitarian disasters. In the Kasai region of the Democratic Republic of the Congo, over 7.7 million people are at risk of not having access to enough nutritious food. In Borne, Yobe and Adamawa in north-east Nigeria almost 3 million people are facing hunger. Since South Sudan gained independence in 2011, approximately 60 per cent of the population has suffered from the effects of famine. The WFP is assisting with these crises and also providing support for people affected by hunger caused by civil war in Syria and Yemen. Overall, the WFP's disaster response has helped more than 80 million people worldwide.

6.4.2 Who gives food aid?

The major donor countries to the WFP in 2018 are shown in **TABLE 1**.

TABLE 1 Major funding contributors to the WFP in 2018 (US$)

All donors and funding sources		
1	USA	2 541 479 166
2	European Commission	1 113 106 906
3	Germany	849 141 329
4	United Kingdom	617 188 873
5	Saudi Arabia	247 907 959
6	United Arab Emirates	226 215 581
7	Canada	222 172 109
8	UN Other Funds and Agencies (excl. CERF)	151 703 536
9	Sweden	148 185 097
10	UN CERF	138 632 047

Figures current as at 28 April 2019

6.4.3 CASE STUDY: Plumpy'Nut — a short-term solution to malnutrition

In 2005 a revolutionary approach to treating malnutrition was released. This is a ready-to-use therapeutic food (RUTF) called Plumpy'Nut. It is a sweet, edible paste made of peanut butter, vegetable oils, powdered milk, sugar, vitamins and minerals.

Its advantages are that it:
- is easy to prepare
- is cheap (a sachet costs about $2.50, including shipping costs)
- needs no cooking, refrigeration or added water
- has a shelf life of two years.

Children suffering from malnutrition can be fed at home without having to go to hospital. It is specially formulated to help malnourished children regain body weight quickly, because malnutrition leads to stunting of growth, brain impairment, frailty and attention deficit disorder in children under two years of age.

Plumpy'Nut is not a miracle cure for hunger or malnutrition; it only treats extreme food deprivation, mainly associated with famines and conflicts. It is not designed to reduce chronic hunger resulting from long-term poor diets or malnutrition. Since its introduction, Plumpy'Nut has lowered mortality rates during famines in Malawi, Niger and Somalia.

FIGURE 2 Plumpy'Nut benefits children.

Most of the world's peanuts are grown in developing countries, where allergies to them are relatively uncommon. Manufacturing plants have been established in a dozen developing countries, including Mali, Niger and Ethiopia. These factories provide employment and ensure ease of access when needed. The patent for Plumpy'Nut is owned by the French company Nutriset. Nutriset has worked with UNICEF to save the lives of millions of children with this simple solution to childhood hunger.

6.4.4 CASE STUDY: Cash vouchers and school feeding programs

Where food is available but people simply cannot afford to buy it, aid is given by the WFP in the form of cash vouchers, which can be exchanged for food and other essential commodities. They allow recipients greater choice in the types of food and other commodities they can obtain. Cash has benefits for local economies because the money is spent within the community. Recently, cash voucher programs have been enhanced through the use of mobile phones, which have been used to provide instant payments to both beneficiaries and the shopkeepers who honour vouchers.

Another program provides schoolchildren with either full meals (breakfast and/or lunch) or nutritional snacks, such as high-energy biscuits. In some cases, school meals are provided alongside take-home rations that benefit the whole family and provide an added incentive for sending children to school. In 2017, the WFP provided meals to 18.3 million school students.

Australia has funded school feeding programs in Bangladesh, Myanmar, Laos and Cambodia, which have had strong positive impacts on both the children and the wider community. School rates of enrolment have increased and regular attendance has improved. Households have also benefited through a reduced need to purchase food. In 2017–18, Australia provided over $108 million to the WFP. This included $38 million in core funding, $2 million to provide school meals and $68 million towards disaster relief.

DISCUSS

Discuss the issues that may arise as a consequence of a country deciding to slash its overseas food aid program by half.   **[Ethical Capability]**

 Resources

🔗 **Weblink** World Food Programme

6.4 INQUIRY ACTIVITIES

1. Select a major donor of food aid from **TABLE 1**. Research the main population characteristics of this country, such as life expectancy, literacy levels and death rates. Discuss your findings in class.
Examining, analysing, interpreting
2. Use the **World Food Programme** weblink in the Resources tab to learn about the WFP's involvement in Syria and surrounding *places* since 2012. What action is the WFP taking there and why?
Examining, analysing, interpreting
3. Draw a poster or advertisement to inform Australians about Plumpy'Nut and its uses and impacts.
Classifying, organising, constructing

6.4 EXERCISES

Geographical skills key: GS1 Remembering and understanding **GS2** Describing and explaining **GS3** Comparing and contrasting **GS4** Classifying, organising, constructing **GS5** Examining, analysing, interpreting **GS6** Evaluating, predicting, proposing

6.4 Exercise 1: Check your understanding

1. **GS1** Refer to **FIGURE 1**. How many countries receive WFP assistance each year?
2. **GS1** Refer to **FIGURE 1**. Where does the majority of the WFP's work take place? Why is this so?
3. **GS1** How much did Australia contribute to the WFP in 2017–18, and how was this distributed?
4. **GS2** Explain why the WFP is so active in school feeding and emergency aid programs.
5. **GS2** Refer to the case study 'Cash vouchers and school feeding programs'. List the advantages and disadvantages of cash vouchers.

6.4 Exercise 2: Apply your understanding

1. **GS6** How might food aid *change* when a donor country experiences a major economic downturn?
2. **GS6** Predict the likely consequences for children who suffer from malnutrition. Present your information in an appropriate diagram.
3. **GS6** Should Australia's food aid commitment be increased? Write a letter to your federal member of parliament, outlining your views on increasing Australia's food aid contribution.
4. **GS2** Suggest advantages and disadvantages of using Plumpy'Nut or other RUTFs to treat childhood malnutrition in developing countries.
5. **GS5** In 2018, how much more did the USA contribute to the WFP than the next highest single country donor? Suggest reasons as to why this might be so.

Try these questions in learnON for instant, corrective feedback. Go to www.jacplus.com.au.

6.5 SkillBuilder: Constructing a box scattergram

online only

What is a box scattergram?

A box scattergram is a table with columns and rows, which displays the relationship between two sets of data that have been mapped. The distribution becomes clear, although in a generalised way, as there are usually only four to five categories of data.

Select your learnON format to access:

- an overview of the skill and its application in Geography (Tell me)
- a video and a step-by-step process to explain the skill (Show me)
- an activity and interactivity for you to practise the skill (Let me do it)
- questions to consolidate your understanding of the skill.

Hunger level (% undernourished)	Aid received per person (US$)			
	No data	Less than 20	20–99	Over 100
35+				• Congo • Mozambique
25–34			• Chad • Angola	
15–24				
5–14		• Nigeria	• Niger	
Less than 5		• South Africa • Algeria • Libya		
No data				• Mauritania

Resources

Video eLesson Constructing a box scattergram (eles-1734)

Interactivity Constructing a box scattergram (int-3352)

6.6 SkillBuilder: Constructing and describing proportional circles on maps

What are proportional circle maps?

Proportional circle maps are maps that incorporate circles, drawn to scale, to represent data for particular places. Different-sized circles on a map reflect different values or amounts of the particular factor being studied.

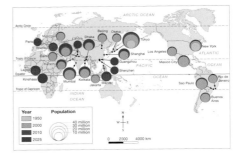

Select your learnON format to access:

- an overview of the skill and its application in Geography (Tell me)
- a video and a step-by-step process to explain the skill (Show me)
- an activity and interactivity for you to practise the skill (Let me do it)
- questions to consolidate your understanding of the skill.

6.7 Do Australians need food aid?

6.7.1 Who is in need?

In 2018, over 3 million people, or 13.2 per cent of Australians, were living below the internationally accepted poverty line. This included almost 739 000 children, or 17.3 per cent of the total Australian child population. The prices of essentials — food, health, education, housing, utilities and transport — have climbed so much in recent years that people who were already struggling are now unable to cope. They may need food aid. The economic climate has seen people turning to charity who in the past would never have dreamed of seeking such support. It is not just traditionally vulnerable groups, such as the homeless, who are seeking food relief; it is also the aged, single parents and the 'working poor'.

In 2018 it was reported that:

- one in eight Australian adults were living in poverty
- one in six Australian children were living in poverty
- many people were living up to $135 per week below the poverty line
- most people (53 per cent) facing poverty were receiving Newstart, Youth Allowance or other government allowance payments
- the biggest threat to household finances were housing costs, including rent
- women made up 52 per cent of adults living under the poverty line.

For many people, charity food agencies are a vital source for their daily food needs. Programs such as the School Breakfast Clubs, managed by Foodbank Victoria in partnership with the Victorian government, provide free, nutritious breakfasts to children who might otherwise start their day without food. Other charities such as OzHarvest, The Big Umbrella, Fareshare and SecondBite work to redistribute food that would otherwise go to waste, providing millions of meals for the many thousands of people who through economic disadvantage may be unable to regularly provide for their own needs.

6.7.2 CASE STUDY: SecondBite

SecondBite rescues and redistributes food to agencies that service people in need. Food is donated directly from farms, as well as from wholesalers, markets, supermarkets and caterers.

SecondBite was founded in Victoria in 2005. Run by just three volunteers, in that year they redistributed 600 kg of food. Since these humble beginnings, SecondBite has grown dramatically. It is now a national organisation, operating with over 85 staff and partnering with more than 1300 community food programs to deliver food and meals to people in need. In the 2015–16 financial year alone, SecondBite rescued and redistributed, free of charge, enough food to provide 20 million meals.

FIGURE 1 SecondBite redistributes food to agencies that assist people in need.

6.7.3 CASE STUDY: APY lands food security plan

In September 2011, Indigenous Australian peoples in South Australia's far north faced food insecurity. Shops in the Anangu Pitjantjatjara Yankunytjatjara (APY) lands were reasonably well stocked, but people were undernourished because of the high cost of freighted fresh food. Essential foods in remote community stores were more than double the price of those in Adelaide. To alleviate the situation, the Red Cross and the South Australian government sent pallets of food to aid impoverished people living in the APY lands.

A government-developed food security plan for the area was established in 2011, with a focus on implementing improvements to food supply (through measures such as improved freight efficiency, stores management, cold storage upgrades and provision of generators for more reliable power supply), community education in choosing and preparing nutritious foods, and an arid lands horticulture project to develop capacity to produce fresh food within the region.

Since 2014, food security measures within the region have largely been driven by non-governmental organisations (NGOs) and the Mai Wiru Regional Stores Council Aboriginal Corporation, which coordinates and manages a number of the community stores within the region and uses store profits to fund various community projects. Improvements have been made, but the challenges of providing affordable, healthy fresh food to remote areas such as the APY lands are ongoing.

FIGURE 2 Indigenous communities living in remote areas face challenges to food security due to the cost of transporting and storing fresh food.

6.7.4 CASE STUDY: Meals on Wheels

As Australia's population ages (see **FIGURES 3** and **4**), the services of groups such as Meals on Wheels may also be in greater demand. In 1997, the **median age** was 34 years, but this is projected to be 44–46 years in 2050. In 1997, people aged 65 years and over comprised 12 per cent of the population, and this is projected to rise to 24–26 per cent in 2050.

FIGURE 3 Australia's population pyramid, 2016

FIGURE 3 Australia's population pyramid, 2016

FIGURE 4 Australia's projected population pyramid, 2050

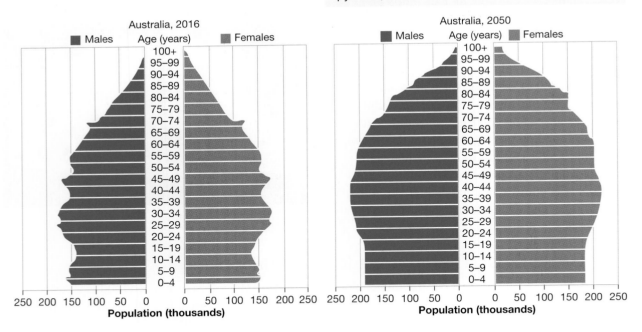

Meals on Wheels began in the United Kingdom during World War II, and in Australia (Melbourne) in 1952. Through delivering nutritious, relatively inexpensive meals, (a three-course meal generally costs between $7 and $10), Meals on Wheels plays an important role in helping older people and those living with a disability to live independently. Around 15 million meals are served annually to more than 50 000 people across the nation. In addition to providing vital nourishment to those who may have difficulty in preparing their own meals at particular stages of life, the social interaction provided by these regular visits is another important aspect of this service.

on Resources

🔗 **Weblink** Australian poverty

6.7 INQUIRY ACTIVITIES

1. Conduct your own research into a local organisation that provides food aid. Use a PowerPoint presentation to detail your findings.   **Examining, analysing, interpreting**
2. Use the **Australian poverty** weblink in the Resources tab to discover other aspects of poverty in Australia.
                                Examining, analysing, interpreting

6.7 EXERCISES

Geographical skills key: GS1 Remembering and understanding **GS2** Describing and explaining **GS3** Comparing and contrasting **GS4** Classifying, organising, constructing **GS5** Examining, analysing, interpreting **GS6** Evaluating, predicting, proposing

6.7 Exercise 1: Check your understanding

1. **GS1** Examine **FIGURES 3** and **4**.
   (a) How many Australians were over 65 years of age in 2016?
   (b) How many are expected to be over 65 years in 2050?
2. **GS1** How many people were living in poverty in Australia in 2018?
3. **GS1** SecondBite redistributes enough food to provide how many meals each year?

4. **GS2** Explain the importance of volunteers in food redistribution.
5. **GS2** Explain how Australia's size could lead to food shortage in some *places*.

6.7 Exercise 2: Apply your understanding

1. **GS2** Explain why there might be difficulties with access to food in 2050 if 25 per cent of the population is over 65.
2. **GS6** Predict whether Meals on Wheels will experience an increase or decrease in its future clientele. Apart from the ageing population, what other factors might *change* the demand for services such as SecondBite or Meals on Wheels in future?
3. **GS6** What would be your family's reaction if the cost of food doubled because of freight costs? What steps could improve the situation in outback areas?
4. **GS6** Have your attitudes to redistributing food *changed* as a result of your reading and class discussion? Explain.
5. **GS6** 'When bills have to be paid, food becomes a **discretionary item**.' (Food Bank Australia 2011.)
   If household bills have to be paid before buying food, what are the likely consequences for families and organisations supplying food aid?

Try these questions in learnON for instant, corrective feedback. Go to www.jacplus.com.au.

6.8 The effects of dietary changes on food supply

6.8.1 How have diets changed?

The human diet has changed throughout history, and continues to change today. Since the 1960s, the total calories per day consumed globally, together with the proportion of the diet comprised of animal products, oils and sweeteners have increased. These food types are typically found in higher amounts in the **Western-style diet** eaten by much of the population of developed countries. **FIGURE 1** shows the changing global diet, as recorded in a study of dietary trends from 1961 to 2009. **TABLE 1** presents the data for each food category.

Since 1960, diets around the world have become more similar and larger in terms of calories, protein, fat and food weight. While animal products, oils and sweeteners have long been a feature of the diet in developed countries, they are increasingly becoming part of the diet in developing countries also. These trends are predicted to continue along with these countries' economic development (see **FIGURE 2**). This is especially the case in countries such as India and China, where the standard of living is rising and people can increasingly afford access to a wider variety of foods.

FIGURE 1 Changes to global diet 1961–2009

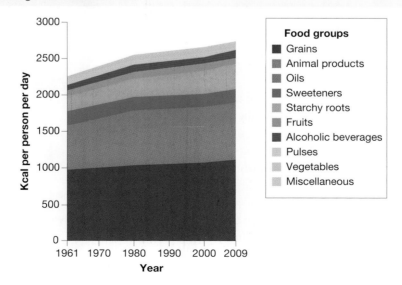

Food groups
- Grains
- Animal products
- Oils
- Sweeteners
- Starchy roots
- Fruits
- Alcoholic beverages
- Pulses
- Vegetables
- Miscellaneous

TABLE 1 Global diet calories per person per day, by food category, 1961–2009

Food category	1961	1980	2000	2009
Grains	976	1042	1077	1118
Animal products	383	473	488	508
Sweeteners	220	275	277	281
Starchy roots	214	184	180	178
Oils	186	255	316	349
Fruits	85.6	94.8	101	104
Alcoholic beverages	79.3	93.6	85	88.2
Pulses	63.4	61.3	58.9	64.9
Vegetables	31.8	40.7	53	55.4
Miscellaneous	0.8	3.4	5.8	9.1
Total calories per day	2239.9	2522.8	2641.7	2755.6

FIGURE 2 Changing diets in developing countries

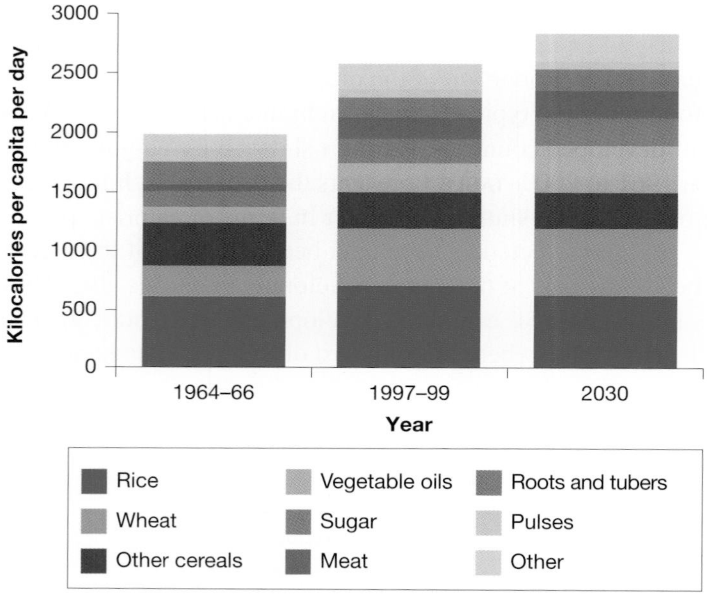

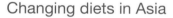

3000
2500
2000
1500
1000
500
0
Kilocalories per capita per day
1964–66
1997–99
2030
Year
Rice
Wheat
Other cereals
Vegetable oils
Sugar
Meat
Roots and tubers
Pulses
Other

FIGURE 3 Impacts of economic growth and dietary change

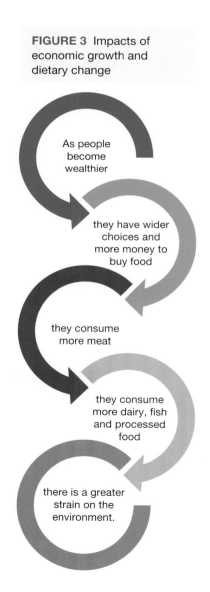

As people become wealthier
they have wider choices and more money to buy food
they consume more meat
they consume more dairy, fish and processed food
there is a greater strain on the environment.

Changing diets in Asia

For centuries, the typical Chinese diet was rice and vegetables, supplemented by fish and small amounts of other meat. Rice is a valuable source of protein, but as people's incomes grow, per capita rice consumption is expected to decline and consumption of protein via meat sources is expected to increase accordingly.

Australians and Americans are the world's highest consumers of meat, eating an average of around 300 grams per person per day — significantly more than the global average of around 115 grams per person. In 1962, the average Chinese person ate just 4 kilograms of meat per year. By 2015, this figure was closer to 80 kilograms (around 220 grams per day) and rising (see **FIGURE 4**).

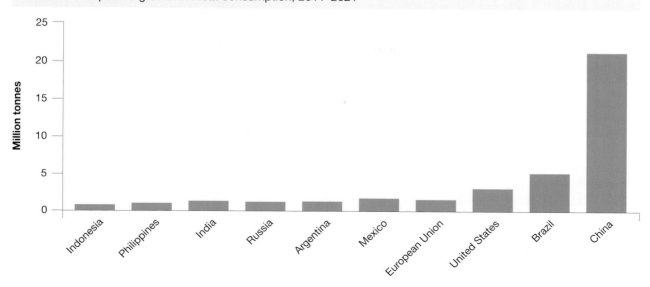

FIGURE 4 Expected growth in meat consumption, 2011–2021

6.8.2 Can Australia be a food bowl for Asia?

The countries of the Asian region are home to more than half the world's population. With significant economic growth occurring throughout much of the area, and over four billion people to be fed, Asia presents unparalleled opportunities for Australian farmers and the Australian economy. As Asian societies become more affluent, higher standards of living give rise to expectations of greater access to goods. Australia is well placed to provide many of these, including a wide variety of foods and quality fibres such as wool and cotton.

With a reputation for utilising 'clean and green' agricultural systems, coupled with our geographic proximity, Australian farmers are ideally placed to capitalise on the economic opportunities that the fast-developing Asian region presents.

'We have the potential for a new golden era of Australian agriculture, given the rise of Asia,' our prime minister said in 2012. The challenge for Australian farmers will be in meeting this booming global need for food and fibre by increasing production at a time when we have decreasing arable land, less water and fewer people working in agriculture.

6.8.3 How can dietary change enable sustainable food production?

One-third of the world's grain crop is fed to animals to produce meat. From a sustainability perspective, this can be considered wasteful, as the amount of grain used to feed a cow for the purposes of meat production is 11 times what would be needed to adequately feed a person with grain alone. Similarly, while 1500 litres of water are needed to produce 1 kilogram of cereal, 15 000 litres are needed to produce 1 kilogram of meat.

Meeting the needs of future populations is not just the responsibility of farmers and producers. We as consumers can also contribute. Attitudes may need to change towards how and what we eat.

- If we are to feed nine billion people sustainably in 2050, it is unlikely we'll be eating a meat-rich, Western-style diet.
- The world produces enough food to feed 10 billion people. However, a significant portion of our crops is used to feed animals or is used as biofuel to produce energy.
- A switch to a diet containing more plant material would allow land currently used to produce animal feed to instead grow crops to feed humans. Although such a huge change is unlikely, even a small shift can have an impact.

- The Meatless Monday campaign encourages people to go without meat for one day per week. This small change would benefit human health and the health of the planet. Meat production requires a large amount of land, water and energy. Cattle are also the largest source of methane gas, which is one of the main contributors to greenhouse gases.
- It is estimated that there are more than 20 000 edible plants that we do not currently eat. Exploring ways of developing and introducing these into our diets may provide additional, sustainable food sources for future generations. One example of an 'old food' that has become increasingly popular in the modern diet is quinoa (pronounced *keen-wah*). A crop from South America, quinoa was used over 4000 years ago by the Incas. It has high nutritional value and grows in a wide variety of climatic conditions. Another advantage of the crop is that all parts of it can be eaten. Peru and Bolivia supply 99 per cent of the world's quinoa demand, and many other countries are now investigating its suitability for their locations.

Increasing consumption of fruits and vegetables, whole grains, legumes and nuts, and limiting intake from animal sources, fats and sugars will not only have health benefits for individuals but will also benefit the planet, as more land and water resources can be directed to sustainable food crop development.

DISCUSS

'A Western-style diet is going to be unsustainable in the future.' Provide one argument for and one argument against this statement. **[Critical and Creative Thinking Capability]**

 Resources

 Interactivity What are we eating? (int-3331)

6.8 INQUIRY ACTIVITIES

1. How has our diet **changed** over time? Ask your parents, grandparents, and/or other adults you know to describe foods and cooking methods from when they were young. Summarise your findings and share with the class. **Examining, analysing, interpreting**
2. A United Nations report stated that 'As **changing** the eating habits of the world's population will be difficult and slow to achieve, a long campaign must be envisioned, along with incentives to meat producers and consumers to **change** their production and dietary patterns. Healthy eating is not just important for the individual but for the planet as a whole.' Design a television commercial to promote a Meatless Monday campaign. **Classifying, organising, constructing**

6.8 EXERCISES

Geographical skills key: GS1 Remembering and understanding **GS2** Describing and explaining **GS3** Comparing and contrasting **GS4** Classifying, organising, constructing **GS5** Examining, analysing, interpreting **GS6** Evaluating, predicting, proposing

6.8 Exercise 1: Check your understanding

1. **GS2** Refer to **FIGURE 1**.
   (a) Which food category makes up the greatest part of people's diets?
   (b) Which category is the second-largest component?
2. **GS2** Study **TABLE 1**. Between 1961 and 2009, how much has the total daily calorie consumption increased?
3. **GS2** Refer to **FIGURE 2**. What crops are people in developing countries eating more of? Which crops are they eating less of?
4. **GS5** Refer to **FIGURE 3**. What is the **connection** between diet and economic development?
5. **GS2** Why have people's diets **changed** over time?

6.9 Urban farms to feed urban populations

6.9.1 What are the advantages of urban farming?

Farming is usually associated with rural areas, but a growing trend in food production is urban farming. This involves the growing of plants and raising of animals within and around cities, often in unused spaces — even the rooftops of buildings.

In many industrialised countries, it takes over four times more energy to move food from the farm to the plate than is used in the farming practice itself. Properly managed, urban agriculture can turn urban waste (from humans and animals) and urban waste water into resources, rather than sources of serious pollution. In 2000, about 15 to 20 per cent of the world's food supply came from urban gardens; in 2018, more than 800 million people practised urban agriculture, contributing to over 20 per cent of all global agricultural production.

Benefits of urban farming include:

- increasing the amount, variety and freshness of vegetables and meat available to people in cities through sustainable production methods
- improving community spirit through community participation, often including disadvantaged people
- incorporating exercise and a better diet into people's lives, leading to improved physical and mental health
- using urban waste water as a resource for irrigation, rather than allowing it to become a source of serious pollution
- reducing the percentage of income people spend on food.

Urban farming could become more important with rapid urbanisation. With the developing countries in Africa, Asia and Latin America expected to be home to 75 per cent of all urban dwellers, they will face the problems of providing enough food and disposing of urban waste.

6.9.2 CASE STUDY: Kolkata sewage ponds

The East Kolkata wetlands in India (see **FIGURE 1**) cover 12 500 hectares and contain sewage farms, pig farms, vegetable fields, rice paddies and over 300 fishponds. With a population of more than 14 million, the **urban agglomeration** of Kolkata produces huge volumes of sewage daily. The wetlands system treats this sewage, and the nutrients contained in the waste water then sustain the fishponds and agriculture. About one-third of the city's daily fish supplies come from the wetlands, which are the world's largest system for converting waste into consumable products. The wetlands are also a protected **Ramsar site** for migratory birds. However, the area is now under pressure from urban growth and from the subsequent increase in waste that it needs to treat.

FIGURE 1 Catching fish in the Kolkata wetland system fishponds

6.9.3 CASE STUDY: Container fish farming

On a smaller scale, a German company has developed a sustainable form of aquaculture that can be used in small spaces in cities. It is called **aquaponics**. Fish swim in large tanks in a recycled shipping container (see **FIGURE 2**). Electric pumps move the fish-waste-filled, ammonia-rich water into a **hydroponic** vegetable garden in a greenhouse mounted above the tank. The fish waste fertilises tomatoes, salad leaves and herbs growing in the greenhouse, and the plants purify the water, which is returned to the tanks.

These structures can be set up almost anywhere, such as on rooftops and in car parks, and the sustainably produced fresh vegetables and fish can be delivered to nearby city markets and shops, reducing the distance that the products must travel. Farmers only need to feed the fish and keep the fish-tank water topped up to sustain the efficient aquaponic system.

FIGURE 2 Urban farming — fish and agriculture

 **Resources**

🔗 **Weblinks**   Urban aquaponics
                 Vertical farming
🛰 **Google Earth** Kolkata

6.9 INQUIRY ACTIVITIES

1. Use the **Urban aquaponics** weblink in the Resources tab to outline the advantages of aquaponics.
   Examining, analysing, interpreting
2. Use the **Vertical farming** weblink in the Resources tab to help you understand vertical farming.
   (a) Draw an annotated diagram to illustrate vertical farming.
   (b) Research an urban farming project in a city. Present your findings in a PowerPoint presentation.
   Classifying, organising, constructing

6.9 EXERCISES

Geographical skills key: GS1 Remembering and understanding **GS2** Describing and explaining **GS3** Comparing and contrasting **GS4** Classifying, organising, constructing **GS5** Examining, analysing, interpreting **GS6** Evaluating, predicting, proposing

6.9 Exercise 1: Check your understanding

1. **GS1** What are the main features of urban farming?
2. **GS1** What functions do the East Kolkata wetlands perform?
3. **GS1** How do communities benefit from urban farms?
4. **GS2** What is hydroponic gardening?
5. **GS2** Outline how an aquaponic gardening system works.

6.9 Exercise 2: Apply your understanding

1. **GS6** Consider the idea of vertical farming.
   (a) Predict the **places** in the world likely to have vertical farms.
   (b) Explain why you selected these **places**.
2. **GS6** Think about urban farming. Could urban farms encourage agricultural tourism? Explain your view.
3. **GS6** Write a letter to the minister for planning, suggesting that urban farming **spaces** should be included in every new urban development.
4. **GS5** Suggest what the advantages and disadvantages might be of producing food on the rooftop **spaces** of city buildings. What factors might influence the types of food that could be produced on rooftops?
5. **GS5** When investigating urban farms and people's gardening activities in Denver, United States, researchers found that:
   - people's community pride improved
   - graffiti and vandalism decreased
   - gardeners felt a greater **connection** with their local **place**.
   Are these worthwhile results from urban farming? Explain.

Try these questions in learnON for instant, corrective feedback. Go to www.jacplus.com.au.

6.10 Thinking Big research project: Community garden design

SCENARIO

Urban sprawl affects food production areas. One solution is to create urban farms. You will design (and perhaps build!) a community garden for your school or neighbourhood, to produce fresh vegetables and fruit to sell locally or that might be used in your school canteen.

Select your learnON format to access:

- the full project scenario
- details of the project task
- resources to guide your project work
- an assessment rubric.

 Resources

💡 **ProjectsPLUS** Thinking Big research project: Community garden design (pro-0192)

6.11 Review

6.11.1 Key knowledge summary
Use this dot point summary to review the content covered in this topic.

6.11.2 Reflection
Reflect on your learning using the activities and resources provided.

 Resources

> **eWorkbook** Reflection (doc-31722)
>
> Crossword (doc-31723)
>
> **Interactivity** Meeting our future global food needs crossword (int-7648)

KEY TERMS

anthropogenic resulting from human activity (man-made)

aquaponics a sustainable food production system in which waste produced by fish or other aquatic animals supplies the nutrients for plants, which in turn purify the water

discretionary item an item that is bought out of choice, according to one's judgement

genetically modified describes seeds, crops or foods whose DNA has been altered by genetic engineering techniques

hydroponic describes a method of growing plants using mineral nutrients, in water, without soil

median age the age that is in the middle of a population's age range, dividing a population into two numerically equal groups

Ramsar site a wetland of international importance, as defined by the Ramsar Convention — an intergovernmental treaty on the protection and sustainable use of wetlands

urban agglomeration the extended built-up area of a place, including suburbs and continuous urban area

Western-style diet eating pattern common in developed countries, with high amounts of red meat, sugar, high-fat foods, refined grains, dairy products, high-sugar drinks and processed foods

yield gap the gap between a certain crop's average yield and its maximum potential yield

UNIT 2
GEOGRAPHIES OF INTERCONNECTION

Every text, call, purchase or trip we make connects us to information, other people and places. This interconnection is influenced by people's views or perceptions of these places. Our consumption of goods and services and our travel, recreational and cultural choices all have impacts on the environment. This has implications for future sustainability.

FIELDWORK INQUIRY: WHAT ARE THE EFFECTS OF TRAVEL IN THE LOCAL COMMUNITY?

online only

Your task

Your team has been commissioned by the local council to compile a report evaluating the impacts of travel movements around a local school or traffic hotspot. You will need to collect, process and analyse suitable data and then devise a plan to better manage future traffic and pedestrian movement in the area.

Select your learnON format to access:
- an overview of the project task
- details of the inquiry process
- resources to guide your inquiry
- an assessment rubric.

on Resources

💡 **ProjectsPLUS** Fieldwork inquiry: What are the effects of travel in the local community? (pro-0149)

7 Connecting with our places

7.1 Overview

Exciting or dull, familiar or strange? How can the same place look and feel different for each person?

7.1.1 Introduction

Geography is the study of people and their connections with places. The way we interact with places is dynamic: we change places and places change us. In a world of nearly eight billion people, we have many different perceptions of what a place is like, how it is used and how it could be improved. More people are on the move, too. Their journeys may be on foot or by plane as they visit and interact with new places. With rapid developments in technology, some of those places may be imagined. What do our connections look like today, and how will they change tomorrow?

On Resources

☑ **eWorkbook**   Customisable worksheets for this topic

▤ **Video eLesson** Making connections (eles-1722)

LEARNING SEQUENCE

7.1 Overview
7.2 'Seeing' places
7.3 **SkillBuilder:** Interpreting topological maps   `online only`
7.4 The meaning of 'land'
7.5 Changing places
7.6 Modes of accessing places
7.7 Walking to connect
7.8 **SkillBuilder:** Constructing and describing isoline maps   `online only`
7.9 Providing access to places for everyone
7.10 Connecting with the world
7.11 **Thinking Big research project:** Fieldwork — moving around our spaces   `online only`
7.12 **Review**   `online only`

To access a pre-test and starter questions and receive immediate, **corrective feedback** and **sample responses** to every question, select your learnON format at www.jacplus.com.au.

7.2 'Seeing' places

7.2.1 Perceptions of places

People's **perceptions** of places are rarely the same. A person's view of a particular place or region is coloured by their own culture, experiences and values. The characteristics and significance of a place will be viewed differently by each individual, and our mental maps of the world can change daily as we have new experiences and gain new knowledge.

The biggest influences on the way we perceive places are age, gender, class, language, **ethnicity**, race, religion and values. How important a place is to us may be determined by whether we feel that place belongs to us or not, whether it is part of our tradition or history, or whether the place is totally unfamiliar.

A place can seem exciting, scary, interesting or boring depending on our experience, expectations or mood on a particular day. Our perceptions of places may also change over time according to climatic changes, conflict or economic shifts.

It is important to understand the factors that influence our perceptions of places and regions, as well as the impact that other groups and cultures have on our perceptions. If we can understand those influences, we may be able to avoid the dangers of **stereotypes** and appreciate the diversity that exists around us.

FIGURE 1 The Kaaba in Mecca — sacred, interesting or crowded?

7.2.2 Mapping places

We all form an impression of our physical surroundings — even of places we have never actually been to. These are what geographers call our 'mental maps'.

Mental maps tell us how to order the space around us. There is no such thing as an accurate mental map, but people's mental maps of their immediate environment tend to be more realistic than those of places they have never visited. Think about some of the ways you use mental maps in your daily life. You may direct someone from point A to point B, telling them about landmarks they will see along the way. You may think about the quickest way to get to the city from a friend's house, imagining your route in your mind.

Our mental maps can help document our influences. Those who walk a lot may be more connected with their neighbourhood and surrounding environment, whereas those who drive will have a very different perspective in their mental map. In the 'Streets Ahead' study by VicHealth, children who walked to school drew pictures that included street names and friends' houses, and they were able to describe people and places in detail (see **FIGURE 2a**). Children who were driven to school tended to separate items from their environment, displaying them in distinct windows (see **FIGURE 2b**).

Mental maps of places we are unfamiliar with are heavily influenced by the media and stereotypical discussions. Travel helps to counteract the effects of the media and generally increases a person's knowledge of an area, providing them with a better understanding of what a place is really like.

FIGURE 2 (a) Drawing by a child who walks to school (b) Drawing by a child who is driven to school

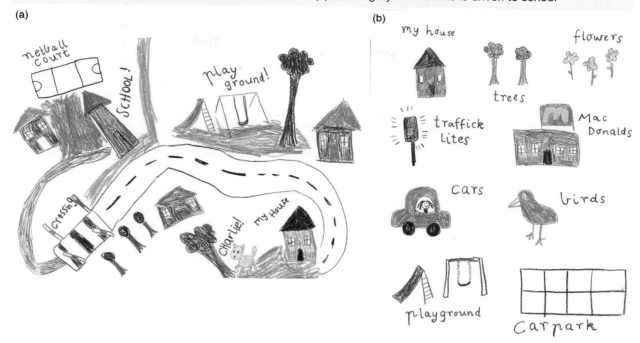

7.2 INQUIRY ACTIVITIES

1. Create a mental map of your journey to school on a blank sheet of A3 or A4 paper. Include as many annotations as you can, such as street names, landmarks, shop names and so on. Once you have finished, compare and contrast the *scale*, size and accuracy of your mental map with a street directory or an ICT mapping tool. Write a paragraph that details some of the differences between your perception and reality.

   Classifying, organising, constructing

2. With your class, make a list of the places or landmarks in your community that you use on a regular basis. Each student should rate the importance of each on a scale of 1 to 3, with 3 being the most important. Collate the data to find out which *places* are most and least important to your class. Are the results as you expected? Do they match your own perceptions of how important *places* are, or do you have a different view from your classmates? Explain why there might be similarities or differences.

   Classifying, organising, constructing

7.2 EXERCISES

Geographical skills key: GS1 Remembering and understanding **GS2** Describing and explaining **GS3** Comparing and contrasting **GS4** Classifying, organising, constructing **GS5** Examining, analysing, interpreting **GS6** Evaluating, predicting, proposing

7.2 Exercise 1: Check your understanding

1. **GS1** List the factors that may influence our perception of *place*.
2. **GS2** From your list in question 1, which factor do you think is the most influential? Why?
3. **GS2** Why do you think the two children's maps shown in **FIGURE 2** are so different? What does it say about their *interconnection* with their *environment*?
4. **GS2** What is the importance of a mental map?
5. **GS2** How does travel influence our mental maps?

▶

7.2 Exercise 2: Apply your understanding

1. **GS6** How do you think the *place* in **FIGURE 1** is viewed by different groups? What kind of experiences or influences may affect their view? Try to provide at least three different possible perceptions for the image.
2. **GS6** The writer Henry Miller once said, 'One's destination is never a *place*, but a new way of seeing things.' What does this quote mean to you, in light of your knowledge of various *places*, your own travels, and what you have learned about perception?
3. **GS2** In what ways do you think your mental map is different for your route to secondary school to what it was for your route to primary school?
4. **GS2** Explain how a significant *place* you have been to (an *environment* or building) altered your sense of *place*.
5. **GS6** Where would you like to travel? Why would you like to go there? What do you expect to see? Do you think the *place* will live up to your expectations?

Try these questions in learnON for instant, corrective feedback. Go to www.jacplus.com.au.

7.3 SkillBuilder: Interpreting topological maps

What is a topological map?

Topological maps are very simple maps, with only the most vital information included. These maps generally use pictures to identify places, are not drawn to scale and give no sense of distance. However, everything is correct in its interconnection to other points.

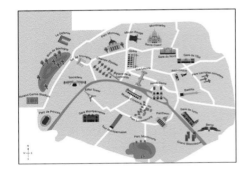

Select your learnON format to access:

- an overview of the skill and its application in Geography (Tell me)
- a video and a step-by-step process to explain the skill (Show me)
- an activity and interactivity for you to practise the skill (Let me do it)
- questions to consolidate your understanding of the skill.

 Resources

Video eLesson Interpreting topological maps (eles-1736)

Interactivity Interpreting topological maps (int-3354)

7.4 The meaning of 'land'

7.4.1 Why is land so important?

Land means different things to different people. A farmer sees land as a means of production and a source of income. A conservationist sees land as a priceless natural resource that must be protected. A property developer sees it as an area that can be divided, built upon and sold for a profit. To Indigenous Australians, land has traditionally been much more than that — it's been an enormously important part of people's culture.

Indigenous Australian peoples have been in Australia since the beginning of the Dreaming (estimated to be more than 60 000 years), adapting to survive and thrive in a changing environment. In traditional Indigenous Australian culture, the land is therefore at the core of people's wellbeing — a person's relationship with the land is one of interconnectedness across the physical, spiritual and cultural worlds (see **FIGURE 1**).

FIGURE 1 A simplified view of Indigenous Australian peoples' relationship with the land

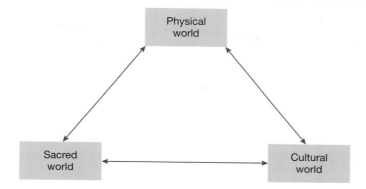

7.4.2 Indigenous Australian peoples' traditional perception of land

In Indigenous Australian culture, land is much more than the soil, rocks, hills and trees. The land, or country, represents the environment that has through history sustained Indigenous Australian peoples, their culture and way of life. Indigenous Australian peoples are diverse, made up of over 500 different groups, each with its own separate language (or dialect), laws, beliefs and customs. Language groups are made up of a number of communities, with each community belonging to a territory or traditional land. These places include features of the natural environment such as waterholes and hills, as well as distinct geographical boundaries such as rivers or mountain ranges. Natural features are often represented in Indigenous art (see **FIGURE 2**).

FIGURE 2 A traditional Aboriginal dot painting depicting land at Kiwirrkura, 400 kilometres west of Alice Springs

Source: © Donkeyman Lee Tjupurrula Kukatja (c.1921)–1994
Tingari Dreaming at Walawala (1989)
Synthetic polymer paint on canvas, 119.7 × 179.3 cm
Purchased from Admission Funds, 1989
National Gallery of Victoria, Melbourne
© Donkeyman Lee Tjupurrula/Licensed by VISCOPY 2013

It is the responsibility of each community to look after their country just as it looks after them. The environment holds rich meaning for Indigenous Australian peoples, whose Dreaming stories (for Aboriginal peoples) and legends (for Torres Strait Islander peoples) are present throughout the landscape, along with many sacred places for special ceremonies — men's and women's sacred sites (see **FIGURE 3**) — and resting places for ancestors that must be protected and conserved.

Each community has a **totem** that is a sign of its people's spiritual link to the land. A totem could be an animal, plant or geographical feature such as a weather pattern or rock formation. It is from this totem or land feature that an individual draws their spirituality, and they feel a special responsibility to protect it. Special ceremonies are performed at sacred places to show respect for, replenish and celebrate each totem.

7.4.3 Differing viewpoints on land

In traditional Indigenous Australian culture, people are custodians rather than owners of land, as the land has existed before and will exist long after its human occupants. It therefore cannot be bought or sold. The concept of property or land ownership that arrived with the Europeans contrasted greatly with the Indigenous view of place.

When the European colonies were established, many Indigenous Australian peoples were dispossessed of their land, and cultural practices were forcibly disrupted. In many cases, Indigenous communities were pushed onto marginal lands that were often not their own, not only creating conflict but severing their connection with the land from which they drew their sense of identity. However, even today among groups largely displaced from their traditional estates, that strong link to country is maintained through stories and a sense of place and spiritual connection.

James Price Point on Western Australia's Kimberley (see **FIGURE 3**) provides one illustration of these differences in viewpoint. The following are three very different views of the same area of land:

- unremarkable beach — Colin Barnett, former premier of Western Australia
- major heritage site — WA Department of Aboriginal Sites ('Major' is the Department's highest category.)
- secret Aboriginal men's business site — Goolarabooloo Aboriginal people.

When different people have vastly different views about a place, it can make the management of that land challenging.

FIGURE 3 James Price Point, Western Australia

7.4.4 Movement across and connection to the land today

Much like the international system of passports and visas to enter other countries, a similar process exists for Indigenous nations. Entry to another nation's or community's lands is by ceremony and negotiation, a practice still commonplace today, recognising the important relationship that Indigenous Australian peoples have with their country. The tradition of 'Welcome to Country' for visitors issues a shared commitment to protect and preserve the land being visited. After being welcomed, those who walk on another's lands are expected to respect the traditional owners' rules and protocols.

The 2016 census showed that 79 per cent of Aboriginal and Torres Strait Islander peoples are living in Australia's urban environments; only 20 per cent live in rural areas. Many of those living in urban environments know the stories passed through generations, but not all have had the opportunity to visit their traditional lands to learn about and experience first-hand their people's particular connection to the land.

DISCUSS

Brainstorm with other members of your class and construct a list of other examples of different cultural viewpoints on the same object, custom or *place*. Consider such things as music, religious customs and foods.

[Intercultural Capability]

7.4 INQUIRY ACTIVITIES

1. Refer to **FIGURE 3** and conduct internet research to find out more about James Price Point and the conflict that developed over the proposed gas mining of the area.
   (a) Create a mind map that shows the various groups involved in the gas mining dispute. Beneath each group's name, list their interests in the site.
   (b) Consider the viewpoints about James Price Point quoted in this subtopic. How have these individuals or groups perceived the land in this *place*?
   (c) The proposed project at James Price Point was cancelled in 2013. From your internet research, why do you think this happened? Share your findings with your classmates. Do they agree?

   Examining, analysing, interpreting
2. Given the strong *interconnection* to land, you may think that Indigenous Australians are opposed to land development. Although custodial responsibilities and care of the land are of utmost importance, many landowners strongly support economic development. As a class or in small groups, debate the arguments in favour of and against development of traditional lands. **Examining, analysing, interpreting**
3. Who are the traditional owners of the land on which you live? Have you witnessed a 'Welcome to Country' ceremony? Who performed the ceremony, and what was involved? (It may have included a speech, traditional dance or smoking ceremony.) **Describing and explaining**

7.4 EXERCISES

Geographical skills key: GS1 Remembering and understanding **GS2** Describing and explaining **GS3** Comparing and contrasting **GS4** Classifying, organising, constructing **GS5** Examining, analysing, interpreting **GS6** Evaluating, predicting, proposing

7.4 Exercise 1: Check your understanding

1. **GS1** How do Indigenous Australian peoples perceive the land?
2. **GS2** What does land mean to you? Think about where you live or where you come from to help describe the *interconnection* you have with the land.
3. **GS1** With their connection to land, what responsibilities does each Indigenous clan have?
4. **GS1** How does each clan's totem connect them to the land?
5. **GS1** What is the difference in viewpoint between being a 'custodian' and being an 'owner' of the land?

7.4 Exercise 2: Apply your understanding

1. **GS5** Describe the natural features of the land depicted in the **FIGURE 2** artwork.
2. **GS2** The establishment of European colonies pushed Indigenous people onto marginal land. How did this *change* their relationship with the land?

3. **GS2** Why do you think the opening of the National Parliament is preceded by a 'Welcome to Country' ceremony?
4. **GS6** Most Indigenous Australians today are urban dwellers. Suggest how these people might connect with the land.
5. **GS5** In 2019 widening of the Western Highway near Ararat in Victoria was realigned in order to save two birthing trees. Why can this be seen as an important decision taken by the Major Road Projects Authority?

Try these questions in learnON for instant, corrective feedback. Go to www.jacplus.com.au.

7.5 Changing places

7.5.1 A chequered history

Places can change very slowly over time and space, or undergo rapid transformations. Melbourne's laneways are an excellent example of how a place once perceived as unsanitary and unsafe is now a thriving and popular part of a metropolis.

During the gold rushes of the 1850s, Melbourne's laneways were well used by people from all walks of life. Then, at the turn of the twentieth century, they began to take a turn for the worse. Criss-crossing the city, their main function was as a place for rubbish disposal. They were dark and dingy, and riddled with disease, crime, gambling houses and brothels. After two world wars, they became home to many immigrants who had nowhere else to live. The city had lost its shine.

Then, in the late twentieth century, something changed. Perhaps influenced by the regeneration that they had seen in European cities on their travels, people began to see the potential of Melbourne's neglected laneways. Small businesses such as art and craft galleries, fashion boutiques and music shops opened. Business owners leased cheap properties in the laneways, away from the main streets with their high rents. Public spaces were regenerated, adding to the city's landscape. Music and entertainment became a reason to go into the city at night.

People are living in the city again, and the CBD is now perceived as a desirable address — in 2017 its resident population was around 45 000, compared with only 700 in the 1980s. The laneways have been part of this revival.

The laneways today

Better lighting, more cleaning and an increased number of people have all contributed to a change in the perception of Melbourne's laneways. **Street art** tours abound, and many laneway bars, cafés and restaurants are desirable places to see and be seen in. The laneways are one of Melbourne's biggest tourist drawcards, and are particularly famous for the vibrant and colourful street art that adorns their walls. Rather than simple **graffiti** or tagging, these are inspiring artworks from some of Australia's (and occasionally the world's) best street artists.

7.5.2 The laneway revival around Australia

The city of Brisbane has also undergone a transformation in the last few years. The Brisbane City Council has overseen the rejuvenation of some of the city's laneways from dingy, unappealing areas to pockets of discovery. Pocket parks have added greenery, and pop-up markets and pop-up restaurants have made the laneways flexible, vibrant spaces. Similarly, in Sydney, laneways have been transformed with art installations and greenery, while in Adelaide, the Laneways Master Plan 2016 has seen the renaissance of its laneways, which are now people-friendly avenues brimming with visitors, small bars and restaurants. The Market to Riverbank Link Project allows people to move, unimpeded and safe, from Adelaide Central Market to the Torrens River.

FIGURE 1 (a) Centre Place, one of Melbourne's revitalised laneways (b) Street art in Duckboard Place, which adjoins ACDC Lane (c) Map of Melbourne's laneways and arcades (d) Named in 2004 as a tribute to iconic Australian rock band AC/DC, ACDC Lane is home to an array of street art and is a popular tourist attraction.

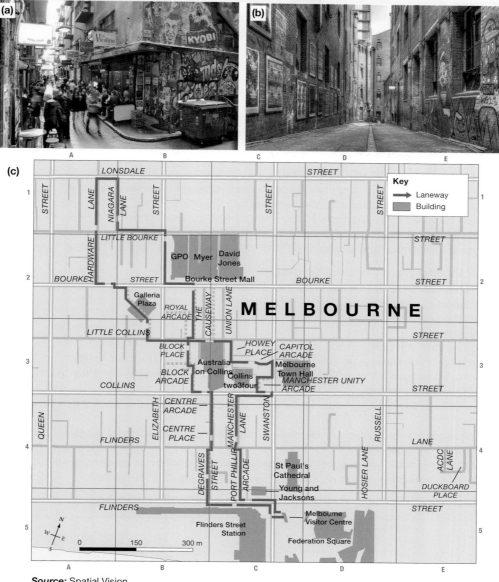

7.5 INQUIRY ACTIVITIES

1. What other areas in Melbourne, or your city, are now developing a laneway culture? How are they attempting to achieve this? In Melbourne, you may wish to investigate Richmond, the QV building or the Docklands precinct. **Examining, analysing, interpreting**
2. Conduct fieldwork in your city's laneways or complete the Laneway Walk shown in the **FIGURE 1(c)** map. Are laneways *sustainable spaces* for people? Give a detailed personal response.
Examining, analysing, interpreting

7.5 EXERCISES

Geographical skills key: GS1 Remembering and understanding **GS2** Describing and explaining **GS3** Comparing and contrasting **GS4** Classifying, organising, constructing **GS5** Examining, analysing, interpreting **GS6** Evaluating, predicting, proposing

7.5 Exercise 1: Check your understanding

1. **GS2** Explain how the perception and uses of Melbourne's laneways have *changed* over time.
2. **GS2** How do laneways allow people to *interconnect* with the city?
3. **GS1** List three *changes* that have made Melbourne's laneways better connected to the city's residents.
4. **GS2** How is street art different to graffiti in the way it connects people with laneways?
5. **GS2** Describe the aspects of the laneways in **FIGURES 1(a)** and **(b)** that connect people to these *places*.

7.5 Exercise 2: Apply your understanding

1. **GS1** What economic factor helped regenerate Melbourne's laneways?
2. **GS6** Some aspects of laneways that can be improved are:
   - waste management and stormwater run-off
   - amenity and access
   - infrastructure, such as public lighting and road surfaces.
   (a) Can you think of any other ways in which laneways can be improved for public use?
   (b) Are there other *spaces* within a CBD *environment* that could be improved in order to provide new *places* for people to enjoy?
   (c) What would need to happen in order to make the *places* you identified in part (b) functional, safe and accessible?
3. **GS6** What *changes* do you think could be made to your neighbourhood shopping area to increase people's connection with this *place*?
4. **GS6** What other uses could you propose for laneway *spaces* in addition to those outlined in this subtopic?
5. **GS6** What role in connecting people do you think laneways may hold in the future?

Try these questions in learnON for instant, corrective feedback. Go to www.jacplus.com.au.

7.6 Modes of accessing places

7.6.1 Connecting with public transport

Public transport provides a relatively low-cost way for people to interconnect with places, and can reduce traffic congestion and pollution. For students, it is often the only way to get around. Sometimes, however, it can seem like too much bother, perhaps because one service does not connect to another or because there are not enough services running, especially near your house.

Public transport use is considerably higher in capital cities than in other parts of Australia, partly because cities have relatively large populations and better public transport **infrastructure**.

Our changing needs

With any population growth, governments at all levels must consider how they will meet changing transport needs. Technology developments have allowed us to make better decisions for our use of public transport. Many people now use the internet or an app to find the fastest way to get from A to B. Service quality, frequency and infrastructure are generally the biggest concerns in the provision of a public transport system. However, the affordability of public transport is equally important, because many people depend upon public transport to access jobs, services, education and recreation.

Different forms of public transport have different uses

Public Transport Victoria aims to deliver quality customer service and provide enhanced access across the range of public transport services throughout Melbourne and Victoria via its website, apps and other forms of social media.

Trains move large numbers of people over long distances at high speed in and out of the central business district (CBD). Greater traveller access is created by routes winding across the city; the fewer stops made by trains and the speed at which they can cover distance increases travellers' ability to connect with places. Melbourne's Metro Tunnel, scheduled for completion in 2025, aims to provide a **turn-up-and-go** experience and increased capacity on the network.

FIGURE 1 Multiple lines facilitate the movement of many trains into and out of central Melbourne.

Trams operate only in areas of high population density, using relatively constant speed and infrequent stops to maximise access for middle-distance commuters.

Buses provide access where trains and trams do not go and 'infill' access for people by using a range of road levels. Buses are the most flexible of the services; they are able to change routes as there is no fixed rail system involved. Buses, and to some extent trams, ferry people to and from train stations, adjusting timetables and reorganising routes to match the train network.

FIGURE 2 Trams take people from the city to suburbs such as Maribyrnong in Melbourne's north-west.

FIGURE 3 Buses and trains interconnect at Ivanhoe station in Melbourne's inner north-east.

7.6.2 User perception of public transport

Conducted quarterly online, the Transport Opinion Survey gathers the views on public transport of 1000 adult Australians. In September 2018, 40 per cent of those surveyed said public transport was a top priority. Only 19 per cent of those surveyed thought Australia's transport systems would be better within a year; just 18 per cent felt that local public transport would be improved within a year; and 34 per cent of those surveyed thought the transport system they were using would be better in five years' time.

A study in the Netherlands revealed that people perceive that their travel time on public transport takes 2.3 times as long as driving a car to make the same journey. People also perceive a continuous journey (involving, say, only one train) as taking less time than a journey that involves transfers and waiting times, even if the second journey is actually shorter. People estimate the waiting time to be about two to three times longer than the actual time. So, a wait of 10 minutes is perceived as 20 to 30 minutes. Factors that influence this perception include:

- uncertainty about when the next bus or train will arrive
- weather conditions
- familiarity with the journey.

Given that travellers tend to consider non-vehicle travel time (walking, waiting, transferring) to be more difficult than in-vehicle travel time, this has consequences when trying to attract people to public transport. If people think their travel time by car is 60 minutes, they perceive their travel time by public transport for that same trip to be almost double: 117 minutes.

7.6.3 Active travel

Cycling and walking to get to work, to visit friends and for recreation have become mainstream modes of transport in the twenty-first century. In particular, in Melbourne's inner suburbs more than 30 per cent of those going to work now choose to use **active travel**.

FIGURE 4 Shared pathways are common.

Melbourne's bicycle paths and trails continue to grow in number, providing increased access to places. **FIGURE 5** maps the 'spiderweb' of pathways around the Melbourne region.

FIGURE 5 Melbourne's bike paths

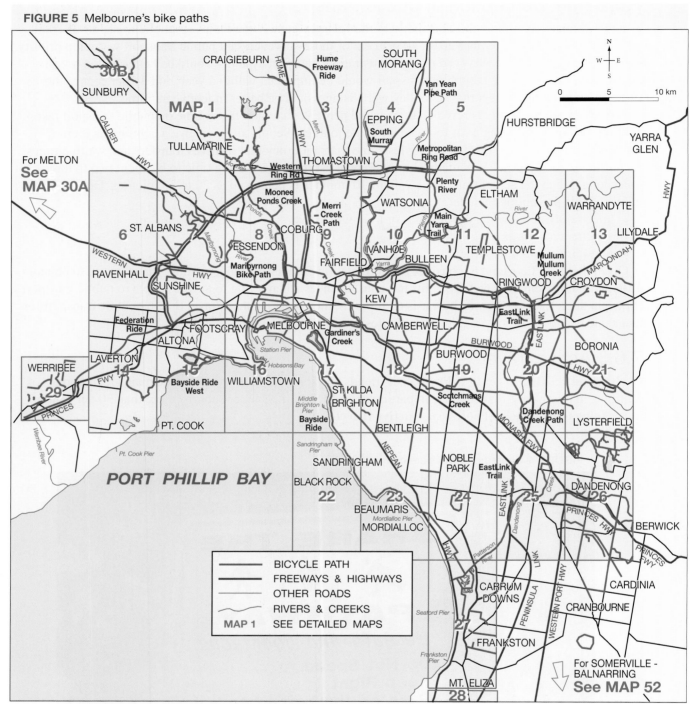

Source: Map courtesy the Bike Paths and Rail trail Guide (Victoria) http://www.bikepaths.com.au

To encourage active travel, railway stations — both new and old — are installing secure bicycle storage areas (called parkiteers; see **FIGURE 6**). Authorities are revising road layouts and regulations to provide a secure riding environment. Manufacturers are also designing electric bicycles to make access available to a wider range of people. Bicycles can be picked up at points within the CBD of many capital cities.

The choice to access places by public transport, active travel or vehicle keeps people connected and strengthens interconnection in a community.

FIGURE 6 Signage at a railway station encouraging the use of parkiteers

 Resources

Interactivity Off the rails (int-3333)

7.6 INQUIRY ACTIVITY

a. Choose a location on the other side of town. Using a rail, bus or other public transport provider website, find out how long it would take you to travel from your school or home to this point on:
   (i) Monday morning at 9 am
   (ii) Sunday evening at 6 pm.
b. What did you notice about the travel times? Were they different? Why do you think this is?
c. Create a map of your journey, using an appropriate key, to show rail, bus and other modes of transport used.

Examining, analysing, interpreting
Classifying, organising, constructing

7.6 EXERCISES

Geographical skills key: GS1 Remembering and understanding **GS2** Describing and explaining **GS3** Comparing and contrasting **GS4** Classifying, organising, constructing **GS5** Examining, analysing, interpreting **GS6** Evaluating, predicting, proposing

7.6 Exercise 1: Check your understanding

1. **GS1** Why is public transport perceived by governments as being very important?
2. **GS1** Write a definition for the term *active travel*.

3. **GS2** Do you use public transport? Why or why not?
   (a) How interconnected is the *place* in which you live? What types of public transport are available to you? What distances do you need to travel to reach a bus stop or a train station?
   (b) What types of public transport are required for you to access your closest international airport?
   (c) How do you perceive the quality of your public transport? Consider accessibility, timeliness, cleanliness, comfort, ticketing, safety, convenience and information about the service. Explain you answer.
4. **GS2** Explain how each form of public transport provides access for different groups in our community. Consider students, workers, senior citizens, those with a disability and tourists. **FIGURES 1, 2** and **3** will provide some additional ideas.
5. **GS2** One of the most significant aspects of public transport is the *interconnection* between the different forms of transport. Why is *interconnection* important?

7.6 Exercise 2: Apply your understanding
1. **GS2** Why do people perceive movement by car as providing preferable connectivity compared to the use of public transport?
2. **GS2** How do you use technology in relation to your transport needs?
3. **GS5** Using **FIGURE 5**, describe the accessibility by bicycle of the following *places* to the Melbourne CBD:
   (a) Frankston
   (b) Altona
   (c) Epping.
4. **GS6** Suggest how the development of electric cars and Uber travel may *change* the way people *interconnect*.
5. **GS6** There are frequent announcements by the Victorian government on developments in the public transport system. For each public transport type and for each form of active travel, suggest two *changes* that might occur in the next ten years. Compare your suggestions with those of others in your class.

Try these questions in learnON for instant, corrective feedback. Go to www.jacplus.com.au.

7.7 Walking to connect

7.7.1 The '20-minute neighbourhood'

Urban planners around the world are focusing on human wellbeing as a key to the structure of new suburbs and revitalisation of existing suburbs. People's perceptions of what will make 'life good' and what makes a 'good place' to live in are being taken into account. Being connected to other places and people is a high priority.

The city of Melbourne is unique, with access to coastal areas, a mild climate, a range of topography, distinctive suburbs or places, considerable tree cover and well-designed buildings and streets. As part of 'Plan Melbourne 2017–2050', the concept of the '20-minute neighbourhood' is currently being implemented.

As **FIGURE 1** shows, the '20-minute neighbourhood' is about improving the liveability of a place. This means being able to walk around your neighbourhood and within 20 minutes being able to access your daily needs — for example transport, a medical clinic, primary schools. Factors that make a good neighbourhood walkable are:
- *a centre* — either as a street or public space
- *people* — enough people for businesses to be successful and for public transport frequency
- *mixed income and mixed use* — a range of housing types
- *parks and public space* — for people to gather and to play
- *pedestrian design* — foot access (cars parked off street)
- *schools and workplaces* — close enough to walk
- *complete streets* — suited to bicycles and walking, and allowing easy movement across the place.

FIGURE 1 The components of the '20-minute neighbourhood' concept

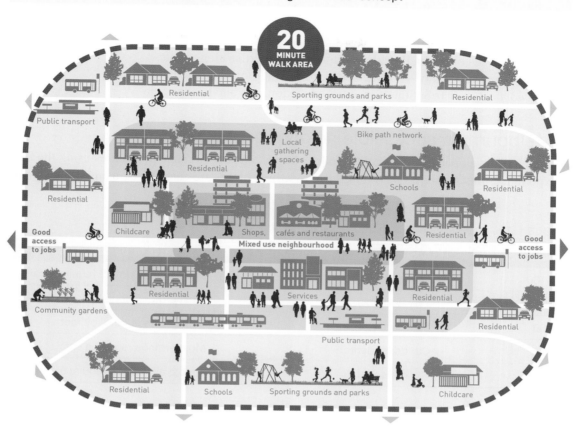

7.7.2 The importance of walkability

Walkability provides a range of benefits to any community. People's health has been shown to improve if they walk on a regular basis. In particular, the risk of heart disease and diabetes is reduced. When people walk regularly, they are often 2.5 to 4.5 kilograms lighter than they would otherwise be. There is a reduced environmental impact with fewer cars on the road: feet produce zero per cent carbon dioxide emissions! Communities benefit when people have more time available for involvement in community activities. Up to 10 per cent of a person's time spent in a community activity is lost when a car is used for just 10 minutes of commuting. Families also benefit financially, because a car is often the second largest household expense, and housing prices can increase by 20 per cent when located in places with a high walkability score.

In Eugene, Oregon, in the United States, areas closest to the centre of the city were shown to have a higher walkability rating than those on the rural–urban fringe (see **FIGURE 2**).

FIGURE 2 Eugene, Oregon walkability ratings, with the most walkable areas shown in red

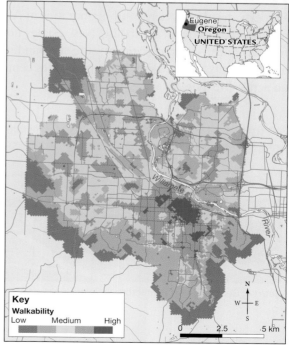

Key
Walkability
Low   Medium   High

0   2.5   5 km

Source: https://www.eugene-or.gov/1229/Full-Composite-Heat-Map

Key elements considered in the mapping and overlaying of Eugene shown in **FIGURE 2** were:
- *density* — the density of the population and the number of employees
- *destinations* — bus stops, shops, primary schools, corner stores, parks and other goods and services
- *distance* — intersection density, bicycle facilities, paths
- *aesthetics* — tree cover, road width, condition of properties along routes
- *safety and perceived safety* — traffic speed, path condition, signalled crossings
- *socioeconomics* — distribution of income, education, age and background.

7.7.3 Melbourne's accessible neighbourhoods

One US company developed the Walkability Index, which considers a range of features using the Eugene, Oregon experience. **TABLE 1** shows the rating scale by which places are categorised using the Walkability Index.

TABLE 1 The classifications within the Walkability Index

Walkability rating	Access to the neighbourhood	Tasks able to be completed
90–100	Walker's paradise	Daily errands do not require a car
70–89	Very walkable	Most errands done by foot
50–69	Somewhat walkable	Some errands done by foot
25–49	Car dependent	Most errands require a car
0–24	Car dependent	Almost all errands require a car

Using this Walkability Index, in 2016 the suburbs of Melbourne that were classified as a walker's paradise (scoring 90–100 on the index rating) included Carlton, Fitzroy, Melbourne CBD, Albert Park, South Yarra, South Melbourne, Collingwood, Southbank and West Melbourne.

7.7.4 Accessible neighbourhoods of the future

Property developers across all major cities in Australia and the developed world have realised the importance of human wellbeing. New estates now focus on providing parklands (often with a water feature); local shopping centres; safe surroundings; foot and bicycle paths; and peaceful, clean, green environments. Advertising for these estates centres on building communities, with young families living an active lifestyle.

FIGURE 3 Development of the Point Cook estate

Source: Central Equity, Featherbrook Point Cook

Planners and developers in established suburbs are seeking to 'infill' the suburbs, creating and re-creating to form 'the 20-minute neighbourhood' around activity hubs and avoid further encroachment on farming land for urban development. The challenge is to have about a 70 per cent increase in housing and population being in the local area hubs. Community and transport infrastructure will need to be revised to achieve this target. In Melbourne, activity hub development can be found at Box Hill, Broadmeadows, Dandenong, Epping, Footscray, Fountain Gate/Narre Warren, Frankston, Ringwood and Sunshine.

FIGURE 4 Box Hill is an important transport hub for trains, trams and buses. There are also a large shopping centre and public and private hospital services in the area. Significant infill development has been undertaken here, with numerous apartment buildings constructed along the major road that runs through the suburb.

7.7 INQUIRY ACTIVITIES

1. Draw a topological map (see subtopic 7.3) of the distances you have to travel from your home to the bus stop or train station, to school, to the shopping centre, to the park where you meet your friends, to a *place* for sporting activities, and to any other significant locations in your life. Discuss in class how teenagers perceive the distances travelled. **Classifying, organising, constructing**

2. Using the internet, search 'How Walk Score Works' and find:
   (a) the walkability rating for Australia's major cities. Comment on their scores.
   (b) the walkability rating for your *place*/home. Can you explain why your *place* has been given its rating?
   (c) the walkability rating of a rural *environment* that you know. Explain why rural areas might be more car dependent. **Examining, analysing, interpreting**

3. On a map of Melbourne, find the suburbs with a high walkability rating. Describe the locations of these *places*. **Examining, analysing, interpreting**

4. In a small group, draw a plan for a 20-minute neighbourhood that you would like to access and live in. Discuss and consider each group member's perception of which features make for wellbeing in a community. **Classifying, organising, constructing [Personal and Social Capability]**

7.7 EXERCISES

Geographical skills key: GS1 Remembering and understanding **GS2** Describing and explaining **GS3** Comparing and contrasting **GS4** Classifying, organising, constructing **GS5** Examining, analysing, interpreting **GS6** Evaluating, predicting, proposing

7.7 Exercise 1: Check your understanding

1. **GS2** How do urban planners use connectivity to make people feel good?
2. **GS1** What is the purpose of a 20-minute neighbourhood?
3. **GS1** Recall and list the features of the 20-minute neighbourhood.
4. **GS2** Using **FIGURE 2**, describe the distribution of the different levels of walkability in Eugene, Oregon.
5. **GS1** What are the benefits of walking for connectivity?

7.7 Exercise 2: Apply your understanding

1. **GS6** Suggest factors that may influence the location of the high level of walkability in Eugene, Oregon.
2. **GS6** To avoid expansion of Eugene:
    (a) Suggest two areas of the city that the city planners and developers might be looking at to improve the level of access. Provide reasons for your choice.
    (b) Suggest a *change* that can be implemented in the short term, medium term and long term to improve access within the city.
3. **GS6** Many parents don't allow their children to walk to school any more. Make a list of the safety issues that parents perceive about access to school.
4. **GS6** Using **TABLE 1**, where would you rank your neighbourhood in terms of walkabiity? Suggest two ways in which your neighbourhood's walkability rating could be raised.
5. **GS6** The 'infill' of current neighbourhoods will make suburbs more connected. How can this be achieved in large cities?

Try these questions in learnON for instant, corrective feedback. Go to www.jacplus.com.au.

7.8 SkillBuilder: Constructing and describing isoline maps

What is an isoline map?

An isoline map shows lines that join all the places with the same value. Isoline maps show gradual change in one type of data over a continuous area. Isolines do not cross or touch each other. The same difference is always shown between each isoline and the next over the entire map.

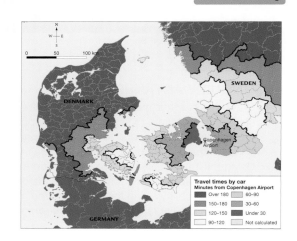

Travel times by car
Minutes from Copenhagen Airport
Over 180 | 60–90
150–180 | 30–60
120–150 | Under 30
90–120 | Not calculated

Select your learnON format to access:

- an overview of the skill and its application in Geography (Tell me)
- a video and a step-by-step process to explain the skill (Show me)
- an activity and interactivity for you to practise the skill (Let me do it)
- questions to consolidate your understanding of the skill.

 Resources

🎞 **Video eLesson**   Constructing and describing isoline maps (eles-1737)

🧩 **Interactivity**   Constructing and describing isoline maps (int-3355)

7.9 Providing access to places for everyone

7.9.1 The challenge of access in cities

Many of us take it for granted that we can walk to the shops, hop on a bus and go to the city centre, or find out when the next train is departing. *Accessibility* refers to people living with a **disability** also having the same access to the physical environment, transportation, information and communication technologies, and other facilities and services. Everyone should feel connected with society, rather than separated from it.

Our cities can be a depressing obstacle course for millions of people. For those living with a disability, negotiating a flight of stairs, opening a door or even reaching a lift button can sometimes be impossible. Have you ever considered how difficult our cities can be for some of their citizens and visitors?

7.9.2 Providing equal access

One in five Australians have reported a disability. In Australia 51 per cent of people over 65 years and 12.5 per cent under 65 report a disability. A 2015 survey revealed that 36 per cent of Australian households include a person living with a disability. Equal access, particularly to transport, is essential for equality. Limiting transport can mean limiting people's opportunities.

FIGURE 1 Percentage of Australians with a reported disability

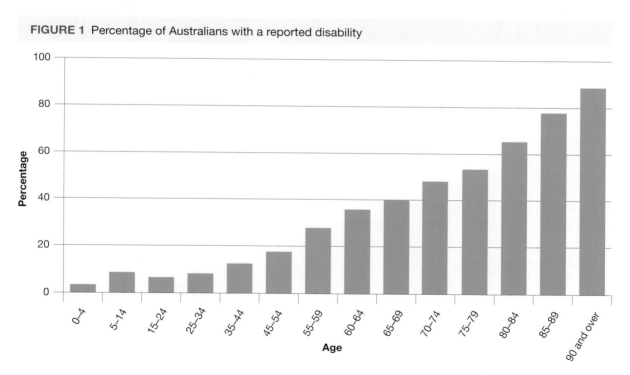

A disability can take many forms, including:
- *walking disabilities* — cannot use stairs easily, moves slowly, needs wider spaces (due to crutches, for example)
- *manipulatory disabilities* — has difficulty in operating handles, vending or ticket machines
- *vision impairment* — has trouble distinguishing between road and pavement, identifying platform edges, knowing whether a lift has arrived at the correct floor, seeing signs or directions
- *hearing problems* — has difficulty hearing announcements about delays, cancellations or emergencies, or hearing an approaching vehicle
- *intellectual disabilities* — is challenged by being in an unfamiliar setting, or coping with cancellations or complex timetables

- *psychiatric disabilities* — experiences stress, anxiety or confusion in crowded situations or encounters with other travellers
- *wheelchair disabilities* — difficulty moving about when no ramps are available, when there are insufficient or badly designed parking spaces, or when there is not enough room to manoeuvre equipment.

There are additional (and sometimes less obvious) disabilities to consider, such as asthma, epilepsy, obesity and diabetes, and temporary disabilities that result from injuries. When considering transport disadvantage, we must also include elderly people, low-income earners, children and outer-urban dwellers, who experience this to some degree as well. Parents with prams or strollers may also be affected.

FIGURE 2 Access for everyone, regardless of mobility, is essential.

7.9.3 A city for everyone

In 2019, the Dutch city of Breda won the European Union's Access City award — a prize for the most accessible European city for people with disabilities. The annual honour aims to award efforts to improve accessibility in the urban environment and to foster equal participation of people living with disabilities. People living with disabilities and their representative organisations were heavily involved in the planning process. Breda won the award for:

- its 'Breda for everyone' municipal website that was tested for its level of accessibility
- making its historic medieval city centre accessible to all, including tourists
- developing the accessibility of its museums, sports centres, theatres, community centres, shops and restaurants
- providing an accessibility fund to assist organisations to make changes to infrastructure
- its commitment to the inclusion of everyone.

Such measures benefit the entire community as well as the economy, because when everyone has access to places of work, people feel included and government social support systems are not overloaded.

7.9.4 Accessing our cities

Victoria is keen to have an inclusive environment for all. 'Absolutely Everyone', the Victorian state disability plan 2017–2020, has four pillars:

- inclusive communities
- health, housing and wellbeing
- fairness and safety
- contributing lives.

Regional cities in Victoria are also conscious of the need to cater for the disabled. The Rural Access Program undertook work in Ballarat to develop a **mobility** map that showed safe, accessible, easy and enjoyable ways to move and connect with the city.

DISCUSS

Investigate your local area to discover to what extent people with disabilities have fair and equitable access to community facilities. (You might like to consider such things as car parking, public transport, access to buildings, ease of movement and safety.) **[Ethical Capability]**

FIGURE 3 Providing mobility within the central business district of Ballarat

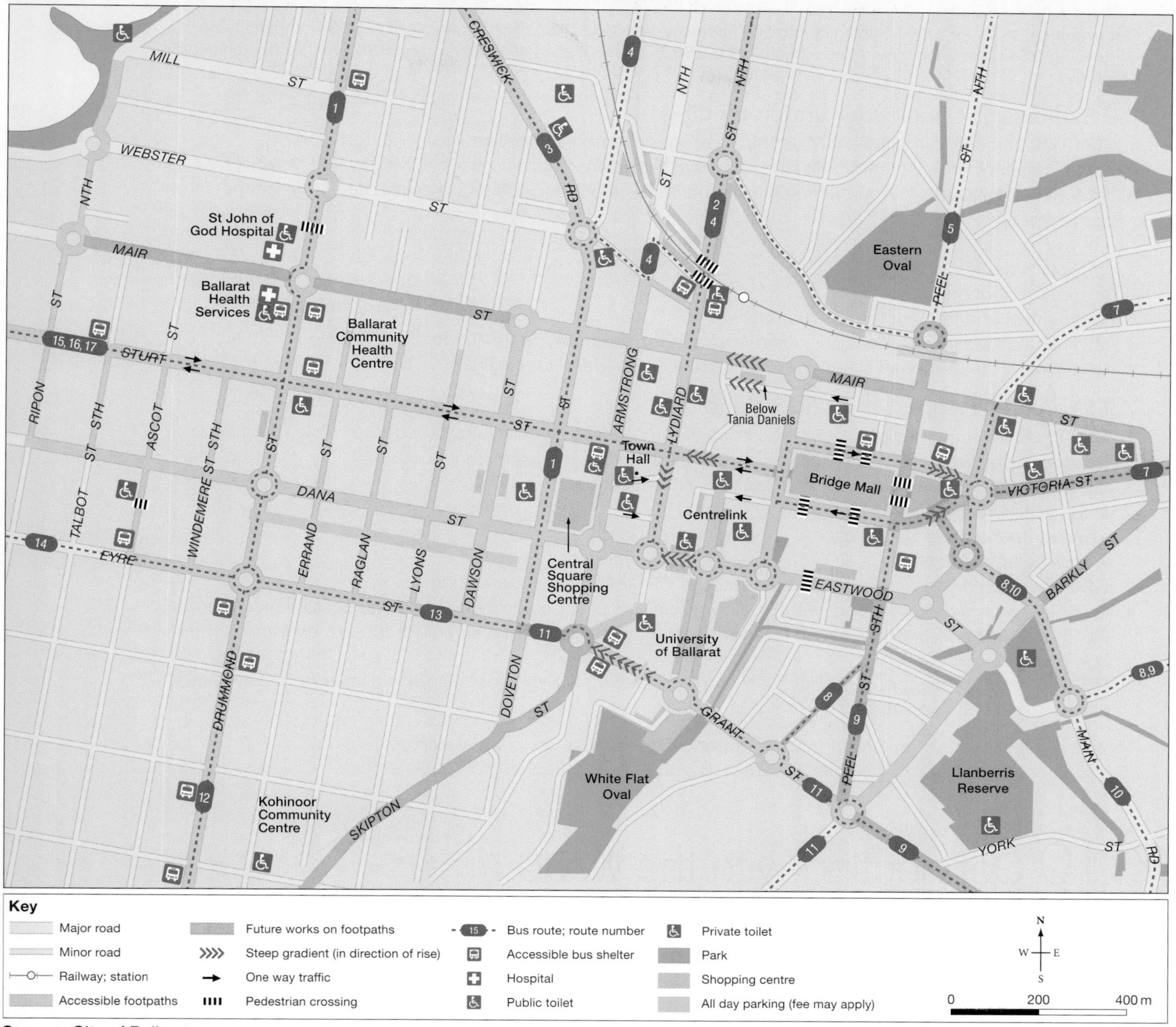
MILL
ST
WEBSTER
MAIR
STURT
DANA
EYRE
ST
NTH
RIPON
TALBOT
STH
ASCOT
WINDEMERE ST STH
DRUMMOND
ERRAND
RAGLAN
LYONS
DAWSON
DOVETON
CRESWICK
RD
ARMSTRONG
LYDIARD
PEEL
VICTORIA ST
EASTWOOD
BARKLY
GRANT
MAIN
YORK
SKIPTON
St John of God Hospital
Ballarat Health Services
Ballarat Community Health Centre
Town Hall
Below Tania Daniels
Bridge Mall
Centrelink
Central Square Shopping Centre
University of Ballarat
Eastern Oval
White Flat Oval
Llanberris Reserve
Kohinoor Community Centre
15, 16, 17
14
13
11
12
Key
Major road
Minor road
Railway; station
Accessible footpaths
Future works on footpaths
Steep gradient (in direction of rise)
One way traffic
Pedestrian crossing
Bus route; route number
Accessible bus shelter
Hospital
Public toilet
Private toilet
Park
Shopping centre
All day parking (fee may apply)
N
W
E
S
0
200
400 m

Source: City of Ballarat

7.9 INQUIRY ACTIVITY

In small groups, devise a trail in a local ***environment*** for people with disabilities. You might choose your school or a local park, for example. You might also decide what types of disabilities you are planning for, and travel around the site considering the potential needs of the visitor. Consider the following:

- Where are hazards located?
- Which areas might the visitor find difficult to navigate?
- What ***places*** might be of interest to them?

Draw an annotated trail map (only for users who are not sight impaired) to highlight these various features. To empathise more fully with the needs of others, students could take it in turns to navigate their way around the designated site on crutches, in a wheelchair or blindfolded. **Classifying, organising, constructing**

7.9 Exercise 1: Check your understanding

1. **GS1** What are the various disabilities that may affect someone's access to public transport?
2. **GS2** Explain what you understand by the term *accessibility* regarding people living with a disability.
3. **GS5** Refer to **FIGURE 1**. Approximately what percentage of people with a reported disability are the same age as:
   (a) you
   (b) your parents
   (c) your grandparents?
   Use figures and percentages in your answer.
4. **GS1** What is the rate of disability within Australia's population? Did this figure surprise you? Explain.
5. **GS1** Why is the European Union's annual Access City Award important?

7.9 Exercise 2: Apply your understanding

1. **GS6** Explain what types of difficulties someone living with a disability may encounter while using public transport. Consider different types of disability and the challenges experienced.
2. **GS5** Using the **FIGURE 3** map of the Ballarat CBD, describe:
   • the accessibility level of Bridge Mall
   • movement around Centrelink
   • travel along Mair Street.
3. **GS6** How would you rank access in the Ballarat CBD? Explain your ranking.
4. **GS6** Rate your neighbourhood (with 5 being 'excellent' and 1 being 'poor') for its accessibility for people living with a disability. Suggest two **changes** that would improve movement for these members of your community.
5. **GS2** Explain how the four key pillars of Victoria's 'Absolutely Everyone' plan could **change** the lives of people living with a disability.

Try these questions in learnON for instant, corrective feedback. Go to www.jacplus.com.au.

7.10 Connecting with the world

7.10.1 How connected are we?

Our world is shrinking. We are more connected than ever before thanks to waterways that have drawn distant places together with improved shipping access, increased and cheaper flights and the digital age. These have brought the world closer to us.

7.10.2 How do maritime highways connect places?

Technological developments have seen the reduction in time for a ship to travel the world. In 2015, the upgraded Panama Canal, which links the Pacific Ocean and the Atlantic Ocean, opened its new, larger locks to accommodate the super ships now plying the oceans. A second Suez Canal lane opened in 2015 and the original canal was deepened to provide access from the Mediterranean Sea to the Red Sea and Indian Ocean. The Straits of Malacca provide access for about 33 per cent of all European container ships accessing East Asia in response to the demand for raw materials and commodities, in particular in China.

Australia is no longer a sailing time of 6 months from the United Kingdom, as it once was; with faster, bigger ships, the distance can be covered in about 33 days. Reduced travel times and reduced costs are a boon for the export of Australian agricultural produce and mining resources, and for the import of products to improve our wellbeing, such as bulky, manufactured goods. Reduced transport costs have benefited global trade.

FIGURE 1 Loading containers for global trade

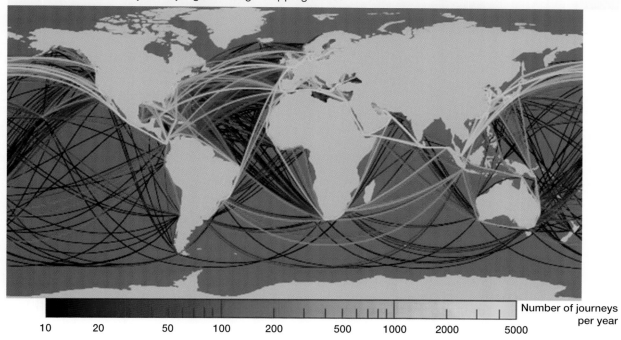

FIGURE 2 The density of major global cargo shipping routes

Number of journeys per year

10 20 50 100 200 500 1000 2000 5000

Cruise ships sail the coastlines of many countries and across local regions, taking tourists on affordable holidays to a variety of places on each voyage. In 2019 passenger numbers of around 30 million people demonstrated the increasing popularity of this mode of travel. Cruise ships continue to grow larger in size, with more berths and a greater range of on-board activities. Access to places has never been easier.

7.10.3 How do we connect through the air?

In 2018, 4.3 billion people flew safely on 38.1 million flights for the purposes of business, tourism or reconnection with relatives. **FIGURE 3** shows how the long-haul aeroplanes, such as the Airbus A380 with its wide body and double-deck carrying capacity of 853 passengers, hub in and out of key airports, leaving smaller jets to distribute passengers across a country using smaller airstrips. These large aeroplanes have reduced the time needed to access places; the Airbus 380 is able to access London from Melbourne in about 22 hours with one stop. Airlines are vying for technologically apt aircraft such as the Dreamliner, which flies long haul, non-stop to London from Perth, Western Australia, in under 17 hours. Constant monitoring of the success levels of routes sees frequent adjustments to schedules and discounts offered on flights. Australians are able to access the United States non-stop (Los Angeles, Dallas) with most flights less than A$1000 each way, and often discounted heavily too.

Air cargo flights also provide access for trade delivering perishable items quickly around the world. The Netherlands trades about 50 per cent of all cut flowers moved around the world. Asparagus from Victoria is sent to Japan, Singapore, Hong Kong, South Korea and Taiwan; it arrives in Japan by air 30 hours after being harvested. Australia imports by air freight high-value 'just-in-time' manufacturing components such as computer and machinery parts. Online shopping can see an order placed in Melbourne via a US site, with the product air-freighted from Hong Kong and delivered in three days!

Technological developments in transport will in the future continue to increase the interconnection of people around the world, making our connections easier, quicker and more frequent.

FIGURE 3 Flight patterns vary according to the time of day.

7.10 INQUIRY ACTIVITY

Use the internet to search 'World traffic pattern over a 24-hour period' and watch major airlines hub into major cities over the course of one day.

a. What do you notice about the places being accessed?
b. Does access change between day and night?
c. Can you explain why international flights leave Australia late at night or early in the morning?

Examining, analysing, interpreting

7.10 EXERCISES

Geographical skills key: GS1 Remembering and understanding **GS2** Describing and explaining **GS3** Comparing and contrasting **GS4** Classifying, organising, constructing **GS5** Examining, analysing, interpreting **GS6** Evaluating, predicting, proposing

7.10 Exercise 1: Check your understanding

1. **GS1** How has the opening of the waterways improved access to places for the cargo ships?
2. **GS2** What is the value of air travel for passengers?
3. **GS2** Refer to **FIGURE 2**. Describe the areas where cargo shipping routes are dense. Is there any connection with the developed world?
4. **GS2** Australia is circumnavigated by shipping routes. Can you explain why this might be?
5. **GS2** What do 'just-in-time' manufacturing components show about global connections?

7.10 Exercise 2: Apply your understanding

1. **GS6** Will Australia's perceived remoteness be further reduced by 2030? Suggest how this might be.
2. **GS2** Describe how a person can connect from Los Angeles, USA to Mildura, Victoria by air.
3. **GS6** International conflicts can have an impact on connecting goods for trade. How might a dispute in the Pacific Ocean affect Australian trade?
4. **GS2** Explain why cruise ship holidays have become so popular.
5. **GS2** Our world is shrinking. Explain this statement in relation to *interconnections*.

Try these questions in learnON for instant, corrective feedback. Go to www.jacplus.com.au.

7.11 Thinking Big research project: Fieldwork — moving around our spaces

SCENARIO

You have been selected as part of a student task force to investigate the problems created by movement of students in and around the school canteen throughout the day. You will conduct fieldwork in the school grounds to assess student movement patterns, identify impacts and formulate possible solutions.

Select your learnON format to access:

- the full project scenario
- details of the project task
- resources to guide your project work
- an assessment rubric

Resources

 projectsPLUS Thinking Big research project: Fieldwork — moving around our spaces (pro-0193)

7.12 Review

7.12.1 Key knowledge summary
Use this dot point summary to review the content covered in this topic.

7.12.2 Reflection
Reflect on your learning using the activities and resources provided.

on Resources

eWorkbook Reflection (doc-31724)

Crossword (doc-31725)

Interactivity Connecting with our places crossword (int-7649)

KEY TERMS

active travel making journeys via physically active means such as cycling or walking

disability a functional limitation in an individual, caused by physical, mental or sensory impairment

ethnicity cultural factors such as nationality, culture, ancestry, language and beliefs

graffiti the marking of another person's property without permission; it can include tags, stencils and murals

infrastructure the facilities, services and installations needed for a society to function, such as transportation and communications systems, water pipes and power lines

mobility the ability to move or be moved freely and easily

perception the process by which people translate sensory input into a view of the world around them

stereotype widely held but oversimplified idea of a type of person or thing

street art artistic work done with permission from both the person who owns the property on which the work is being done and the local council

totem an animal, plant, landscape feature or weather pattern that identifies an individual's connection to the land

turn-up-and-go frequent and regular transport service such that reference to a timetable is not required; e.g. users know that a train will run every 10 minutes

8 Tourists on the move

8.1 Overview

Can a simple relaxing holiday really create jobs, support an economy or affect the environment?

8.1.1 Introduction

For Australians in the 1950s and 1960s, overseas travel was an exotic, time-consuming and expensive adventure that for many was simply beyond their reach. Fast forward to 2020 and nearly 60 per cent of the population now owns a passport. Whether at home or abroad, travel is an important part of modern life.

The World Tourism Organization estimates that by 2030, five million people will travel each day. Where will these people go and what will influence their choices? What impact will these choices have on the places they visit?

Resources

eWorkbook   Customisable worksheets for this topic

Video eLesson   Moving around (eles-1723)

LEARNING SEQUENCE

8.1 Overview
8.2 The importance of tourism
8.3 Global tourism
8.4 **SkillBuilder:** Constructing and describing a doughnut chart   `online only`
8.5 Australian tourism
8.6 **SkillBuilder:** Creating a survey   `online only`
8.7 **SkillBuilder:** Describing divergence graphs   `online only`
8.8 The impacts of tourism
8.9 Managing environmental impacts — eco-friendly tourism
8.10 Cultural tourism
8.11 Tourism and sport
8.12 **Thinking Big research project:** Design a 7-day cruise adventure   `online only`
8.13 **Review**   `online only`

To access a pre-test and starter questions and receive immediate, **corrective feedback** and **sample responses** to every question, select your learnON format at www.jacplus.com.au.

8.2 The importance of tourism

8.2.1 Defining tourism

The World Tourism Organization defines tourism as the temporary movement of people away from the places where they normally work and live. This movement can be for business, leisure or cultural purposes (see **FIGURE 1**), and it involves a stay of more than 24 hours but less than one year.

FIGURE 1 Purpose of people's travel, 2017

Not specified 6%

Business and professional 14%

Visiting friends and relatives, health, religion, other 27%

Leisure, recreation and holidays 53%

Types of tourist

People travelling for leisure have different interests, reasons for travel and preferred ways of approaching the travel experience. **FIGURE 2** illustrates the location of some of the different types of popular tourist destinations, and **FIGURE 3** identifies some of the key characteristics of different types of tourist, and how they like to travel.

FIGURE 2 Types of tourist destinations

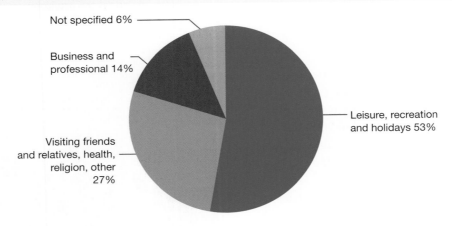

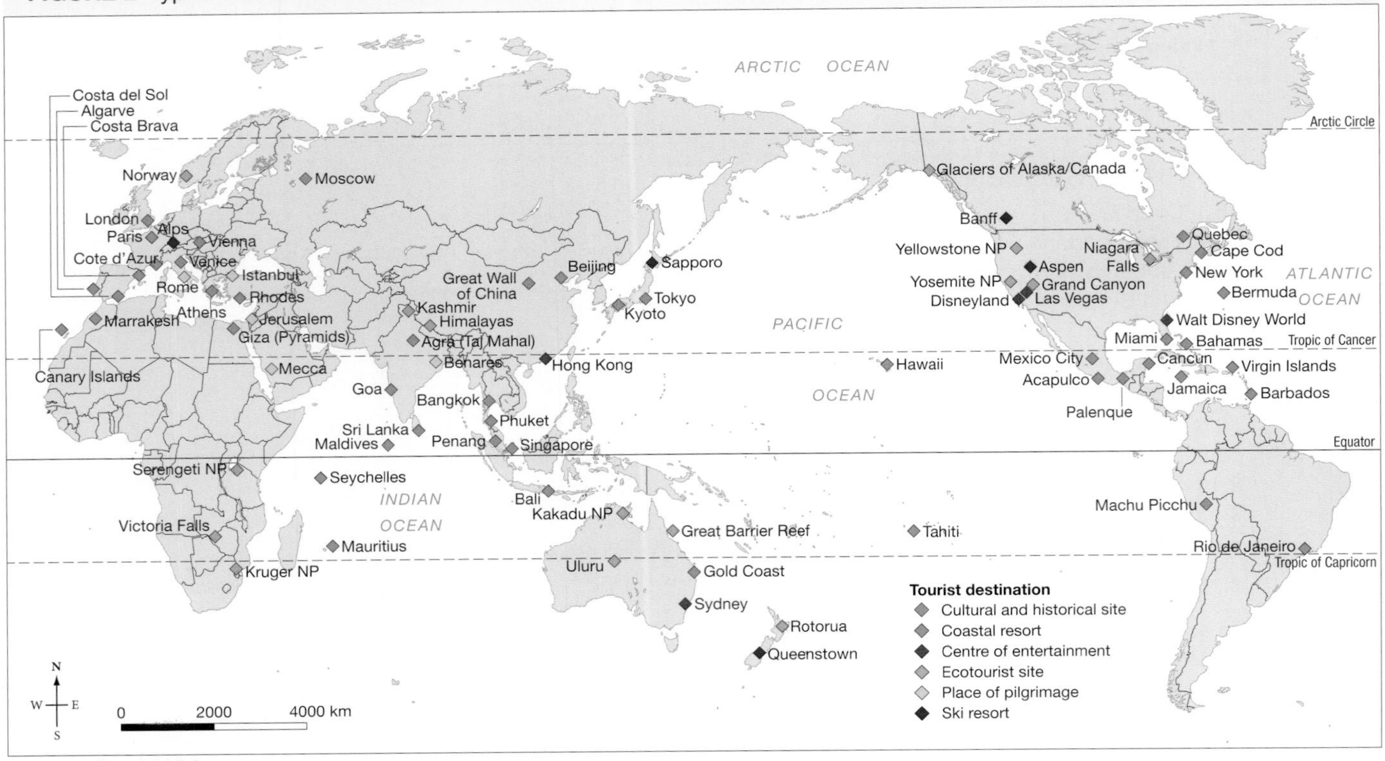

Tourist destination
- Cultural and historical site
- Coastal resort
- Centre of entertainment
- Ecotourist site
- Place of pilgrimage
- Ski resort

Source: Spatial Vision

FIGURE 3 Four kinds of tourist

Organised mass tourist	Individual mass tourist
• Least adventurous • Purchases a package with a fixed itinerary • Does not venture from the hotel complex alone; is divorced from the local community • Makes few decisions about the holiday	• Similar to the organised mass tourist and generally purchases a package • Maintains some control over their itinerary • Uses accommodation as a base and may take side tours or hire a car

The explorer	The drifter
• Arranges their own trip • May go off the beaten track but still wants comfortable accommodation • Is motivated to associate with local communities and may try to speak the local language	• Identifies with local community and may live and work within it • Shuns contact with tourists and tourist hotspots • Takes risks in seeking out new experiences, cultures and places

Medical tourism

Medical tourism involves people travelling to overseas destinations for medical care and procedures. The low cost of travel, advances in technology and lengthy waiting lists caused by the increased demand for elective surgery are turning medical tourism into a multi-billion-dollar industry. In 2017, the global medical tourism market was valued at almost $54 billion, with this figure expected to rise to more than $140 billion by 2025.

While people once travelled overseas only for cosmetic procedures such as facelifts and 'tummy tucks', the range of services offered has expanded dramatically over recent years to include fertility treatments, complex heart surgery, and orthopaedic procedures, such as knee and hip replacements.

Countries all over the world are attracting patients for a variety of reasons. In some instances, it is the high standard of medical care or the outstanding reputation of a particular facility that attracts people, while for others it is the savings to be made and the opportunity to include a holiday and luxury accommodation as part of the package.

Asia is the market leader in the medical tourism industry, with Thailand and India vying for the number one spot. Thailand is slightly more expensive, but offers a better tourist experience and has a wider range of services available. India, on the other hand, is inexpensive and boasts state-of-the-art facilities staffed by Western medical staff, predominantly from the United States. **FIGURE 4** illustrates the savings to be made by having selected medical procedures carried out in Asia rather than in Australia. **FIGURE 5** shows the savings when a variety of procedures are undertaken in Malaysia compared to the United States, Thailand or Singapore. With medical tourism expected to add millions to Asian economies per year, it is not surprising that there has been a dramatic increase in the number of facilities to deliver these services.

FIGURE 4 Cost savings that can be made by having medical treatment in Asia versus Australia

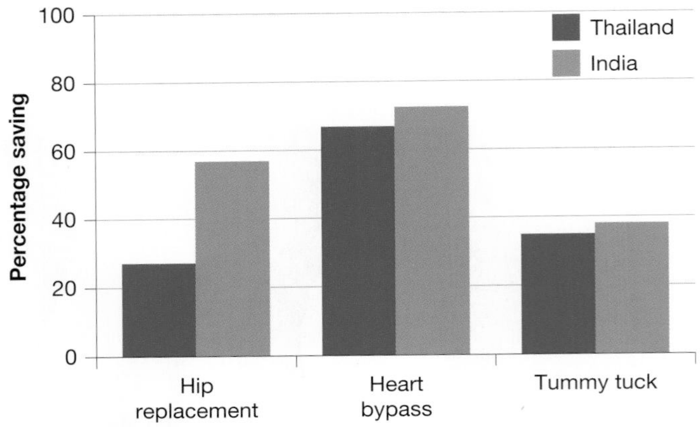

Source: *Adapted from:* Cosmetic Surgeon India and Rowena Ryan/News.com.au

FIGURE 5 Cost savings that can be made by having medical procedures carried out in Malaysia versus the United States, Thailand and Singapore

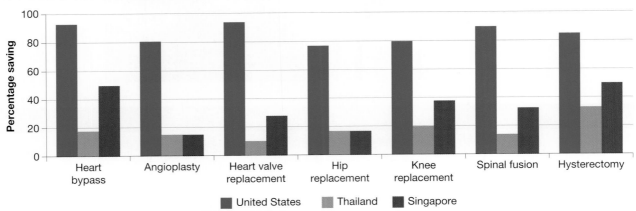

Source: *Adapted from:* Cosmetic Surgeon India and Rowena Ryan/News.com.au

8.2.2 The economic importance of tourism

Tourism is one of the world's largest industries and, as such, an important component in world economies. One in ten jobs worldwide is linked either directly or indirectly to the tourism industry, and in 2017 tourists added US$7 trillion to the global economy. **FIGURE 6** shows the top ten tourism earners for 2017.

Globally, about 10 per cent of **gross domestic product** (GDP) is directly linked to the tourism industry; for many developing countries it is the primary source of income. Even when global economies are experiencing a downturn, people still travel. After natural disasters, countries rely on the return of the tourist dollar to help stimulate their economies.

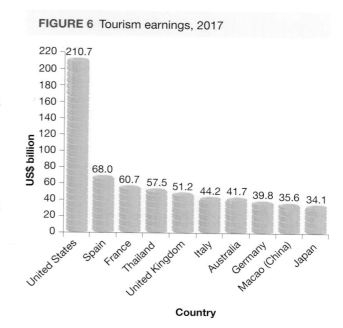

FIGURE 6 Tourism earnings, 2017

Y-axis: US$ billion. X-axis: Country

Values:
- United States: 210.7
- Spain: 68.0
- France: 60.7
- Thailand: 57.5
- United Kingdom: 51.2
- Italy: 44.2
- Australia: 41.7
- Germany: 39.8
- Macao (China): 35.6
- Japan: 34.1

8.2 INQUIRY ACTIVITIES

1. What type of tourist are you? Make a sketch of yourself, similar to the one shown here. Annotate your cartoon to describe yourself as a tourist, using information in this subtopic to help you. Include information about your ideal holiday and explain why you appear as you do in your cartoon. **Classifying, organising, constructing**

2. Using your atlas as a primary source of information, select three *places* from different categories shown in **FIGURE 2** that you might like to visit.
   (a) Calculate the distance between them.
   (b) Explain how you would travel to each *place*.
   (c) Explain what you might expect to see and do in each *place*.
   (d) Work out how long it might take to visit each *place*.
   (e) Describe each location using geographical concepts such as latitude and longitude, direction and *scale*.
   (f) Explain why you have chosen each *place*.

Describing and explaining
Examining, analysing, interpreting

8.2 EXERCISES

Geographical skills key: GS1 Remembering and understanding **GS2** Describing and explaining **GS3** Comparing and contrasting **GS4** Classifying, organising, constructing **GS5** Examining, analysing, interpreting **GS6** Evaluating, predicting, proposing

8.2 Exercise 1: Check your understanding

1. **GS1** What is a tourist?
2. **GS1** What are the two most common purposes of people's travel?
3. **GS1** What are the four different types of tourist? Describe the key characteristics of two of these types.
4. **GS2** What is the main reason for people travelling to Asia for medical procedures?
5. **GS1** Which country had the highest tourism earnings for 2017?

▶

8.3 Global tourism

8.3.1 Who is travelling?

Over time, travel has become faster, easier, cheaper and safer. Economic growth in many parts of the world has ensured that many people now have more money to spend and can afford to travel. Annual leave entitlements provide people with time to travel. For example, in addition to the four weeks annual leave that is a standard condition for full-time employees in Australia, many Australians accumulate long-service leave, which is often spent on an extended overseas trip. It has also become common for young people to spend time seeing the world during a 'gap year' after finishing secondary school, and to travel before establishing a career and perhaps raising a family.

Many young travellers see backpacking as the optimum way to travel. Generally this group:

- is on a tight budget
- wants to mix with other young travellers and local communities
- has a flexible itinerary
- seeks adventure
- is prepared to work while on holiday to extend their stay.

At the other end of the scale, there has also been a dramatic growth in **mature-aged** tourist movements. The number of older people in **developed** countries is growing. Some of these travellers have savings, access to superannuation funds, and the opportunity to retire early; thus, they have both the time and the money to travel.

FIGURE 1 Backpackers tend to travel further and stay longer than other tourists.

8.3.2 Where do people go?

As each tourist enters or leaves a country, they are counted by that country's customs and immigration officials. This data is collected by the World Tourism Organization, and the results can be shown spatially. **FIGURE 2** shows the ten most popular tourist destination countries for 2017.

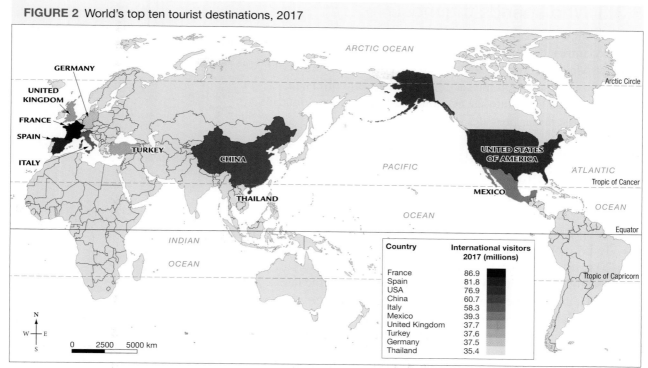

FIGURE 2 World's top ten tourist destinations, 2017

Country	International visitors 2017 (millions)
France	86.9
Spain	81.8
USA	76.9
China	60.7
Italy	58.3
Mexico	39.3
United Kingdom	37.7
Turkey	37.6
Germany	37.5
Thailand	35.4

Source: Data from World Tourism Organization (UNWTO). Map drawn by Spatial Vision.

Where do people stay?

Today, in addition to traditional accommodation such as hotels and backpacker hostels, tourists have a wide range of accommodation options, and their preferences will vary depending on a multitude of personal and economic factors. The rise of operators such as Airbnb means people can now choose to stay independently in an apartment or house, or perhaps in a guest room within someone else's own home. Staying with locals in their homes in cities, towns and villages across the globe provides an opportunity to experience the local culture in a more 'up close and personal' way. For many however, an established resort or hotel remains the preferred choice of accommodation.

When travelling overseas, most tourists give little thought to who owns the hotel or resort in which they are staying. **TABLE 1** lists the locations of various hotel chain headquarters, and shows that the corporate owners of many hotels are based in a country that is often not the one a tourist is visiting.

TABLE 1 World's top ten hotel owners, 2018

Company	Headquarters (country)	Total hotels	Number of countries
Wyndham Hotels & Resorts	USA	8092	66
Choice Hotels International	USA	6429	35
Marriott International	USA	5974	127
InterContinental Hotels Group	UK	5070	100
Hilton Worldwide	USA	4727	104
Accor Hotels	France	4200	100
Best Western Hotels	USA	4196	100
Jin Jiang International	China	3090	67
Home Inns	China	3000	1
Motel 6	USA	1330	2

8.3.3 Who spends the most?

FIGURE 2 shows the countries that attract the most tourists, but which countries do these tourists come from, and how much do they spend? **FIGURE 3** shows the top ten countries in terms of the money they spend on international tourism, and offers an idea of the huge input into the economies of destination countries that the tourist dollar provides.

8.3.4 The growing future of tourism

Year after year, global tourism continues to increase. Advances in transport technology have reduced not only travel times but also cost, making travel increasingly accessible to more and more people.

- Today, you can fly from Australia to Europe in about 20 hours from the east coast, or under 17 hours flying nonstop from Perth. A similar journey by boat in the late 1940s took six weeks or more.
- Airline and tour companies offer a range of cut-price deals, and the increased number of competitors for the tourist dollar means that travel is more affordable.
- Improvements in transport and technology have increased our awareness and knowledge of the world around us and have sparked people's desire to see new places and experience different cultures.
- In general, the travelling public has more leisure time and more disposable income, making both domestic and international travel viable.

FIGURE 4 shows the growth in tourist numbers from 1995 to 2017 and the projected growth through to 2050.

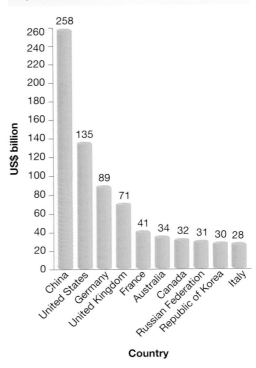

FIGURE 3 World's top ten tourist spenders, 2017

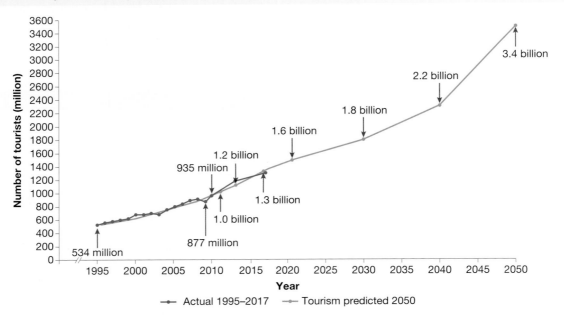

FIGURE 4 Projected future growth in world tourism

The evolving tourist

Improved living standards, increased leisure time and greater disposable incomes have all created opportunities for people to travel and experience new places and cultures. These factors are also shaping the tourist of the future (see **FIGURE 5**). Established and emerging tourist destinations will need to ensure that they meet the evolving needs of the tourist market, in order to continue to attract and benefit from the tourist dollar.

Growth areas for tourism

Predictions suggest that Africa and the Asia-Pacific region will be particular growth areas, attracting more and more tourists in the years to come.

In Africa, for instance, countries such as Kenya and Tanzania offer a different type of tourist experience. Kenya offers:

- relative safety
- beaches and a tropical climate
- safari parks and encounters with lions and elephants
- a unique cultural experience with the **Masai** people.

The influx of tourists to Kenya has led to the establishment of **national parks** to protect endangered wildlife and promote this aspect of the tourism experience. Money flowing into the region can be put towards development projects such as improved water quality and **infrastructure** such as water pipes, roads and airports.

The true challenge for the future, however, is to ensure that:

- money remains in the local economy rather than in the hands of developers, and is used to improve local services, not just tourist services
- the need of indigenous communities to farm the land is balanced with tourist development
- tourist numbers are controlled, to ensure that the environment is not damaged.

Such challenges, of course, are not unique to Kenya. Wherever in the world there is an increase in tourist numbers, there is a need for a sustainable approach, to ensure that the economic benefits of tourism do not come at the cost of a region's people and environment.

FIGURE 5 Characteristics of the future tourist

- Is more mature and experienced
- Has concern for safety and security
- Benefits from increased competition
- Wants value for money
- Seeks experiences more than services
- Adopts new technologies
- Seeks sustainable tourism and development

---

 on Resources

🌏 **Google Earth** Google Earth: Kenya

Explore more with my**World**Atlas

Deepen your understanding of this topic with related case studies and questions.
- Investigate additional topics > Tourism > **World tourism**

8.3 INQUIRY ACTIVITIES

1. (a) On a blank outline map of the world, locate and label the capital cities of each of the top ten tourist destinations.
   (b) Plot a trip from your nearest capital city (e.g. for Victoria, Melbourne) to all ten of these *places*, covering the shortest possible distance, and returning to your capital city. Use the *scale* on the map to estimate the distance travelled.
   (c) Calculate the time it might take to complete this journey.   **Classifying, organising, constructing**

2. The three main types of tourist attraction are natural, cultural and event attractions. Use a dictionary to help you write your own definition of each term. For each of the countries shown in **FIGURE 2**, try to find an example of each type of attraction. Use the **FIGURE 2** map in subtopic 8.2 to help you.

   Classifying, organising, constructing

8.3 EXERCISES

Geographical skills key: GS1 Remembering and understanding **GS2** Describing and explaining **GS3** Comparing and contrasting **GS4** Classifying, organising, constructing **GS5** Examining, analysing, interpreting **GS6** Evaluating, predicting, proposing

8.3 Exercise 1: Check your understanding

1. **GS2** Carefully study **FIGURES 2** and **3** and answer the following.
   (a) On which continents are the top ten destinations located?
   (b) Which continents are generating the most in tourism spending?
   (c) Describe the *interconnection* between destinations and tourism spending.
2. **GS3** What differences in travel needs are there between a mature-age tourist and a backpacker? With the aid of a Venn diagram, show the similarities and differences in the needs of these two groups of tourists.
3. **GS2** Explain why tourism is more accessible to the broader community than it was 100 years ago.
4. **GS2** Explain what national parks are and why they are established.
5. **GS1** How are international tourism numbers calculated?

8.3 Exercise 2: Apply your understanding

1. **GS6** Consider **FIGURE 4**, which shows future growth in world tourism.
   (a) How many global tourists are there predicted to be in 2050?
   (b) Which *places* do you think will be the most popular?
   (c) What impact do you think these increases will have on the *environment*?
   (d) Will this result in small-scale or large-scale *change*?
   (e) Do you think these numbers are *sustainable*? Explain.
2. **GS2** With rapid growth in tourism, there is a need to ensure *sustainability*.
   (a) Explain what you understand by the term *sustainable* tourism.
   (b) Describe an example of tourism that would be considered *sustainable*.
   (c) Describe an example of tourism that would not be considered *sustainable*. Suggest what *changes* might be needed to make it *sustainable*.
3. **GS6** Asia and Africa are future growth areas for tourism; they are also home to many of the world's developing nations. Study **TABLE 1**, which shows hotel ownership. What impact might this ownership have on the countries in which these hotel chains are located?
4. **GS6** Consider the characteristics of the future tourist, shown in **FIGURE 5**. Suggest ways in which each of these characteristics might impact on the tourism choices they make. Which *places* and experiences might be more attractive to this traveller? Which might be less appealing?
5. **GS6** 'Tourists should be able to go where they like and do what they like without any restrictions.' Provide an argument both for and against this statement.

Try these questions in learnON for instant, corrective feedback. Go to www.jacplus.com.au.

8.4 SkillBuilder: Constructing and describing a doughnut chart

What is a doughnut chart?

A doughnut chart is a circular chart with a hole in the middle. Each part of the doughnut is divided as if it were a pie chart with a cut-out. The circle represents the total, or 100 per cent, of whatever is being looked at. The size of the segments is easily seen. Doughnut charts are a useful visual representation of data.

Select your learnON format to access:

- an overview of the skill and its application in Geography (Tell me)
- a video and a step-by-step process to explain the skill (Show me)
- an activity and interactivity for you to practise the skill (Let me do it)
- questions to consolidate your understanding of the skill.

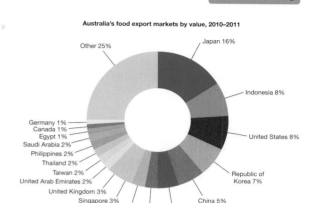

Australia's food export markets by value, 2010–2011

Other 25% · Japan 16% · Indonesia 8% · United States 8% · Republic of Korea 7% · China 5% · New Zealand 5% · Hong Kong 3% · Malaysia 3% · Singapore 3% · United Kingdom 3% · United Arab Emirates 2% · Taiwan 2% · Thailand 2% · Philippines 2% · Saudi Arabia 2% · Egypt 1% · Canada 1% · Germany 1%

on Resources

🎞 **Video eLesson** Constructing and describing a doughnut chart (eles-1738)

🧩 **Interactivity** Constructing and describing a doughnut chart (int-3356)

8.5 Australian tourism

8.5.1 Where in the world are Australians going?

In 2018, 11 million Australians travelled abroad, almost 7 per cent more than in 2017. Of these, some 1.4 million people travelled to New Zealand, making it our most popular tourist destination.

The buying power of the Australian dollar compared to other currencies means that a wide range of international destinations are more affordable than holidaying at home. Competition between airlines, choice of flights and package deals that include combinations of flights, accommodation, tours and meals are largely fuelling the international travel market. The option of children staying for free also makes overseas travel more attractive for families. While over a million people elect to holiday in Australia, for many their tourism dollar has greater buying power in destinations such as Indonesia and Thailand, where the cost of living is much lower than it is at home.

The opportunity to live and work overseas has also seen an increase in the number of people under 30 travelling abroad. The under-30s working visa has ensured that foreign travel is both appealing and affordable for this age group. This visa, which is available in more than 35 countries around the world, allows people aged between 18 and 30 to live and work in a country for up to 12 months. At any one time there are about one million Australians living and working overseas.

While the most popular tourist destination for Australians travelling abroad is New Zealand, the fastest expanding markets for Australian travellers are Indonesia and Japan.

8.5.2 Who comes here and why?

In 2018, 9.1 million tourists came to Australia at a rate of around 1000 per hour. They spent 274 million nights in the country and added $43.9 billion to the Australian economy. The states most visited by international tourists in 2018 were New South Wales, Queensland and Victoria, with Western Australia showing a significant increase in tourist numbers.

It is predicted that by 2050 Australia's tourism industry could grow to be worth $150 billion. In Australia, almost 826 000 jobs can be attributed either directly or indirectly to the tourism industry, representing around 8 per cent of the workforce.

As **FIGURE 1** shows, there are various reasons that people visit Australia, the most popular being simply to holiday. **FIGURE 2** shows the countries of origin of those who visit, and also shows where in the world Australians are travelling on their overseas journeys.

Australia is a land of contrasts, having a wide variety of both human and natural environments. The most popular tourist destinations are shown in **FIGURE 3**.

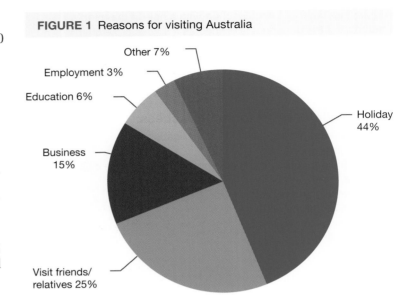

FIGURE 1 Reasons for visiting Australia

- Other 7%
- Employment 3%
- Education 6%
- Business 15%
- Visit friends/relatives 25%
- Holiday 44%

FIGURE 2 Country of origin for tourists visiting Australia, and destinations for Australian tourists

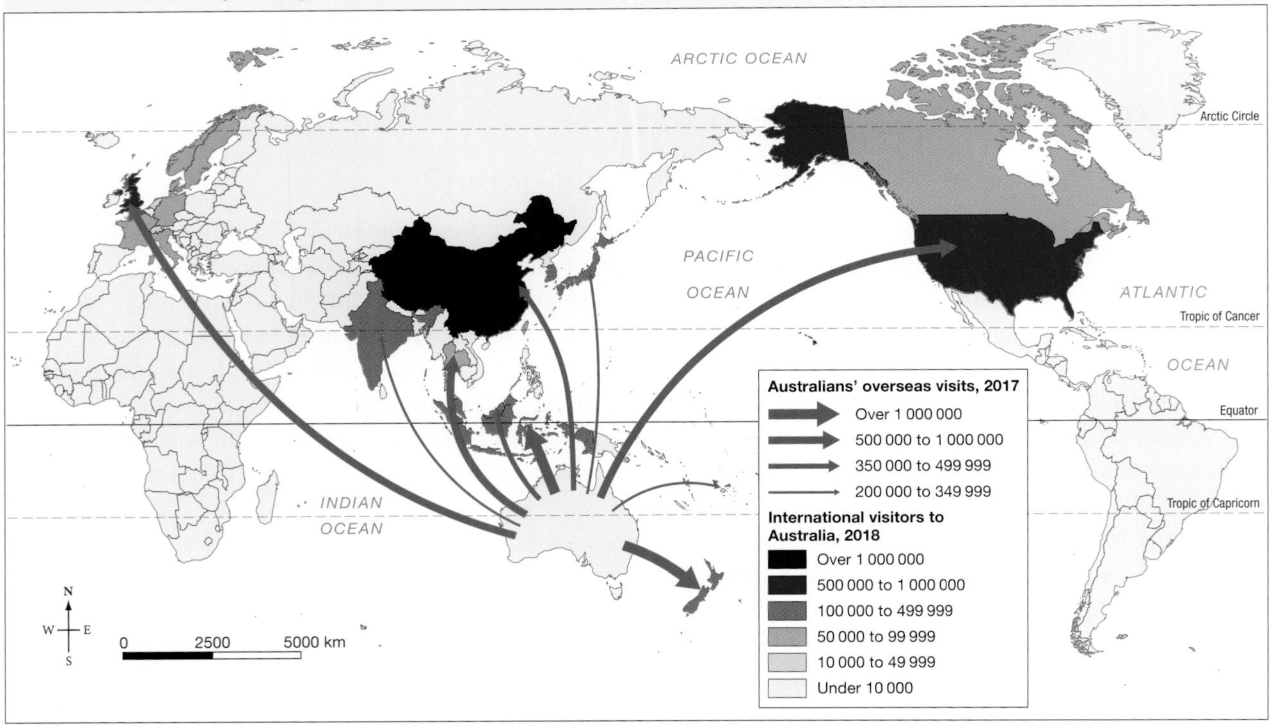

Australians' overseas visits, 2017
- Over 1 000 000
- 500 000 to 1 000 000
- 350 000 to 499 999
- 200 000 to 349 999

International visitors to Australia, 2018
- Over 1 000 000
- 500 000 to 1 000 000
- 100 000 to 499 999
- 50 000 to 99 999
- 10 000 to 49 999
- Under 10 000

ARCTIC OCEAN — PACIFIC OCEAN — ATLANTIC OCEAN — INDIAN OCEAN — Arctic Circle — Tropic of Cancer — Equator — Tropic of Capricorn

0   2500   5000 km

Source: ABS, Austrade

FIGURE 3 Australia's most popular tourist destinations

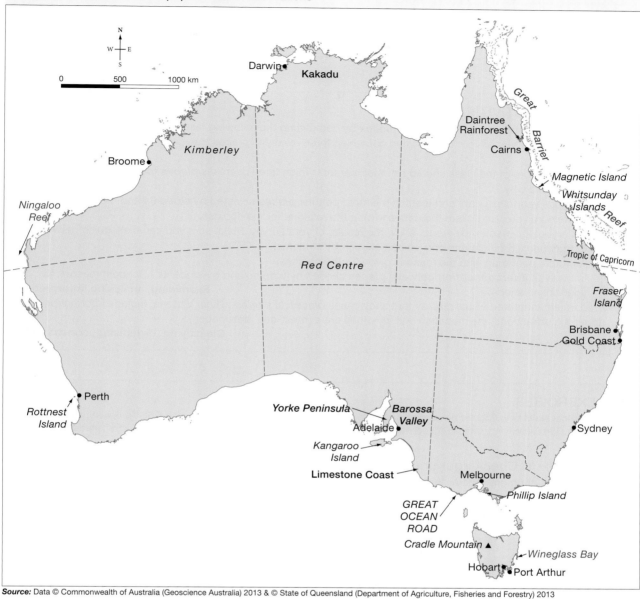

Source: Data © Commonwealth of Australia (Geoscience Australia) 2013 & © State of Queensland (Department of Agriculture, Fisheries and Forestry) 2013

Each Australian state has its own culture and features that attract tourists. **TABLE 1** provides a snapshot of the key feature of the various states. Tourism statistics have revealed that Victoria has increased its share of the international tourist market. Tourism in this state grew by 8.2 per cent in 2018 and added $28.2 billion to the state's economy. Of the 82.3 million visitors Victoria welcomed, 28.4 million stayed overnight, and 54 million were day trippers. New South Wales, however, remains the most visited state, with iconic attractions such as the Sydney Opera House and Sydney Harbour Bridge continuing to draw huge numbers of tourists from all around the world.

TABLE 1 Key attractions of Australian states

New South Wales	Nightlife
Northern Territory	Outback
Queensland	Beaches
South Australia	Wine
Tasmania	Nature
Victoria	Sport
Western Australia	Wine

┌─ Explore more with myWorldAtlas ─────────────────────────────────

Deepen your understanding of this topic with related case studies and questions.
● Investigate additional topics > Tourism > **Tourism in Australia**

8.5 INQUIRY ACTIVITIES

1. Investigate people's favourite overseas *places* by completing the following tasks.
   (a) Survey members of your class to find out which three overseas *places* they would most like to visit and why.
   (b) Each member of the class should ask their parents which three overseas *places* they would most like to visit and why.
   (c) Compile your class data and identify the most popular *places* selected by students and their parents. Make sure you also collate the data showing the reasons for the choices.
   (d) On an outline map of the world, show the results of your survey. Make sure you can distinguish between *places* chosen by parents and *places* chosen by students.
   (e) Annotate your map with the reasons given for the choices.
   (f) Is there an *interconnection* between *places* chosen by parents and by students? Suggest reasons for your observations. **Examining, analysing, interpreting**
2. Prepare an annotated visual display that showcases *places* of interest within Victoria. Include information about the attractions, their location and why they are a 'must-see' destination.
   Classifying, organising, constructing

8.5 EXERCISES

Geographical skills key: GS1 Remembering and understanding **GS2** Describing and explaining **GS3** Comparing and contrasting **GS4** Classifying, organising, constructing **GS5** Examining, analysing, interpreting **GS6** Evaluating, predicting, proposing

8.5 Exercise 1: Check your understanding

1. **GS1** Why might more Australians choose to holiday overseas rather than in Australia?
2. **GS2** Explain the *interconnection* between the *places* most visited by Australians and our major source of tourists.
3. **GS2** Explain the importance of tourism to the Australian economy.
4. **GS5** Which state or territory would you recommend to an international visitor who was interested in each of the following?
   (a) Sporting events
   (b) Experiencing the natural *environment*
   (c) Getting 'off the beaten track'
5. **GS2** Explain what you understand by the term 'day tripper'.

8.5 Exercise 2: Apply your understanding

1. **GS5** Study **FIGURE 3**.
   (a) Identify the top tourist destinations in Victoria and Tasmania.
   (b) Identify which are human and which are natural *environments*.
   (c) Suggest other *places* in Australia that you think should be at the top of every tourist's holiday itinerary.
2. **GS6** Predict the impact on Australian tourism if the Australian dollar was to suddenly lose value in relation to international currencies.
3. **GS6** Predict the impact on Australian tourism if the Australian dollar was to suddenly increase in value and achieve parity with the US dollar.
4. **GS2** What is your favourite tourist destination? What is the appeal of this *place* for you?
5. **GS6** Suggest a strategy for encouraging more Australians to holiday at home and enticing more international visitors to Australia.

Try these questions in learnON for instant, corrective feedback. Go to www.jacplus.com.au.

8.6 SkillBuilder: Creating a survey

What is a survey?

Surveys collect primary data. A survey involves asking questions, recording and collecting responses, and collating and interpreting the number of responses. Because your survey is taken from a relatively small number of people in a population, it is called a sample.

Select your learnON format to access:

- an overview of the skill and its application in Geography (Tell me)
- a video and a step-by-step process to explain the skill (Show me)
- an activity and interactivity for you to practise the skill (Let me do it)
- questions to consolidate your understanding of the skill.

On Resources

| | Video eLesson | Creating a survey (eles-1764) |
| | Interactivity | Creating a survey (int-3382) |

8.7 SkillBuilder: Describing divergence graphs

What is a divergence graph?

A divergence graph is a graph that is drawn above and below a zero line. Those numbers above the line are positive, showing the amount above zero. Negative numbers that are shown indicate that the data has fallen below zero.

Select your learnON format to access:

- an overview of the skill and its application in Geography (Tell me)
- a video and a step-by-step process to explain the skill (Show me)
- an activity and interactivity for you to practise the skill (Let me do it)
- questions to consolidate your understanding of the skill.

On Resources

| | Video eLesson | Describing divergence graphs (eles-1739) |
| | Interactivity | Describing divergence graphs (int-3357) |

8.8 The impacts of tourism

8.8.1 Do the benefits outweigh the costs?

In some ways, tourism seems like the perfect industry. It can encourage greater understanding between people and bring prosperity to communities. However, tourism development can also destroy people's culture and the places in which they live. There is sometimes a fine line between exploitation and sustainable tourism. **FIGURES 2** and **3** outline some of the key positive and negative impacts of tourism.

FIGURE 1 Promoting cultural understanding or commercialising traditional culture? Protecting or exploiting the natural environment? Tourism can have both positive and negative impacts on people and the environment.

FIGURE 2 The positive impacts of tourism

Provides employment and reduces poverty — for example, tourism directly and indirectly supports jobs in Fiji (36.6%) and Belize (37.3%)

Adds more than US$7 trillion to global economies and is the primary source of income for many developing nations. In 2018, Australia's share was $43.9 billion.

To cope with tourist numbers, improvements are made to infrastructure: roads, rail, and sewerage and water supply networks.

Increases interest in the natural environment, so wilderness areas and endangered species are protected. National parks in Kenya promote shooting with cameras, not guns.

Encourages restoration of historic sites, such as Pompeii in Italy and Port Arthur in Tasmania

Preserves indigenous cultures and heritage, because tourist dollars encourage conservation of sites and self-management

Increases employment opportunities for women

Improves health care in poorer communities

Positive impacts of tourism

Increases awareness of the impacts of actions on the environment

Promotes the exchange of cultures and fosters understanding Strengthens cultures, as tourists are interested in the traditions of places they visit, creating a demand for their survival

Generates additional spending — for example, the Chinese government spent US$40 billion on Olympic events, sports facilities, infrastructure, energy, transportation and water supply, all of which benefit the community long after the event

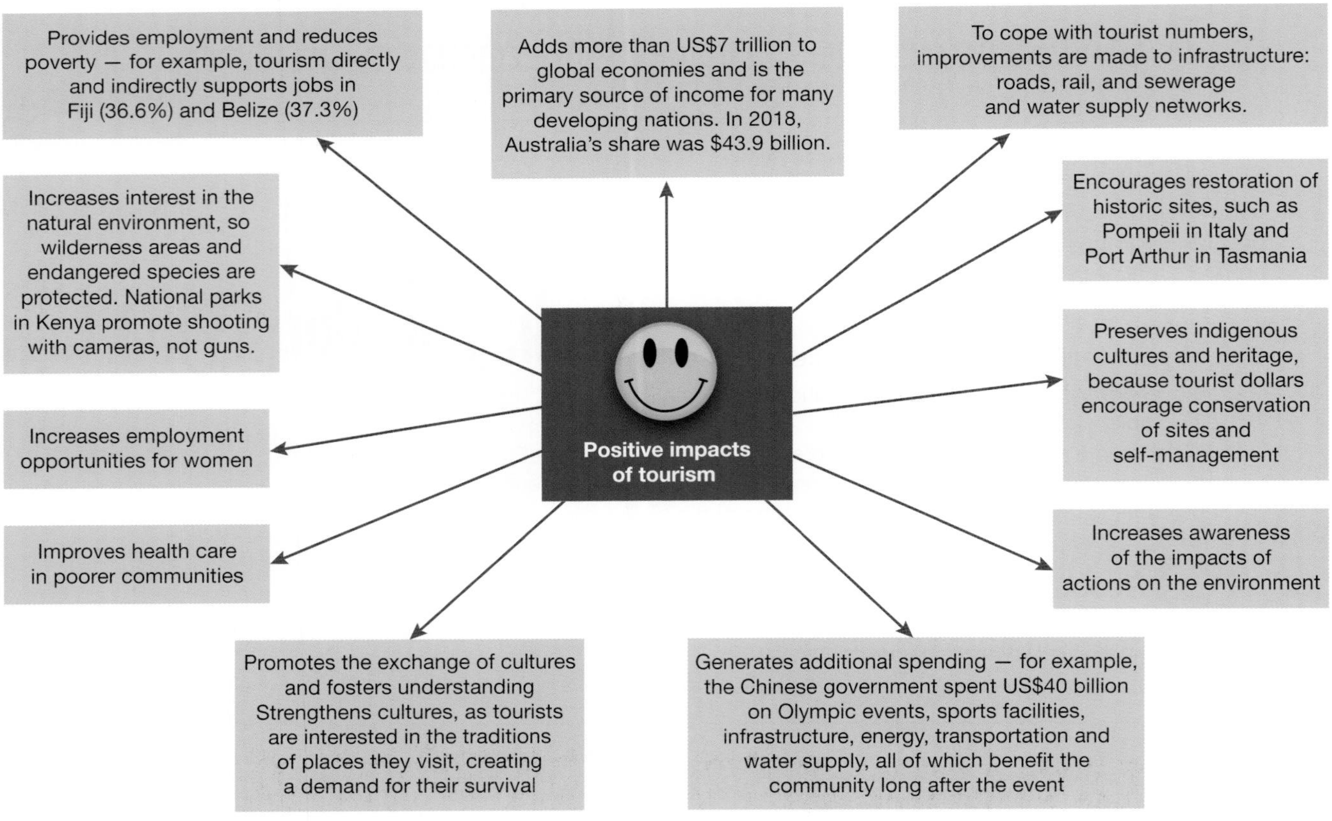

FIGURE 3 The negative impacts of tourism

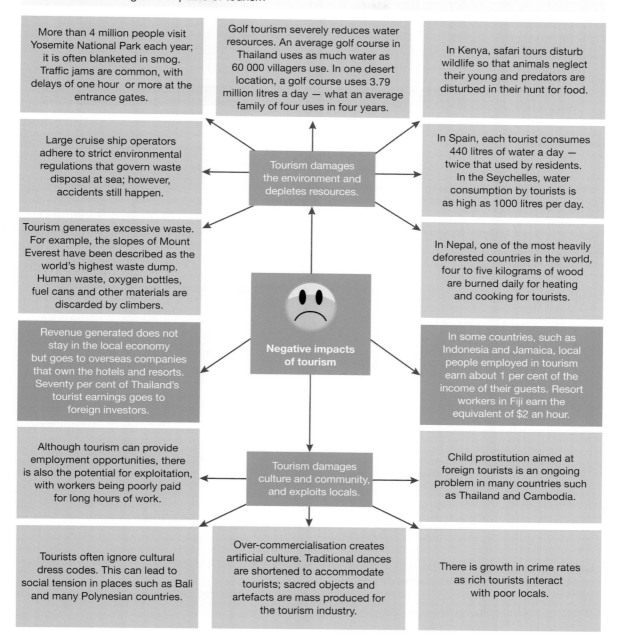

More than 4 million people visit Yosemite National Park each year; it is often blanketed in smog. Traffic jams are common, with delays of one hour or more at the entrance gates.

Golf tourism severely reduces water resources. An average golf course in Thailand uses as much water as 60 000 villagers use. In one desert location, a golf course uses 3.79 million litres a day — what an average family of four uses in four years.

In Kenya, safari tours disturb wildlife so that animals neglect their young and predators are disturbed in their hunt for food.

Large cruise ship operators adhere to strict environmental regulations that govern waste disposal at sea; however, accidents still happen.

Tourism damages the environment and depletes resources.

In Spain, each tourist consumes 440 litres of water a day — twice that used by residents. In the Seychelles, water consumption by tourists is as high as 1000 litres per day.

Tourism generates excessive waste. For example, the slopes of Mount Everest have been described as the world's highest waste dump. Human waste, oxygen bottles, fuel cans and other materials are discarded by climbers.

In Nepal, one of the most heavily deforested countries in the world, four to five kilograms of wood are burned daily for heating and cooking for tourists.

Negative impacts of tourism

Revenue generated does not stay in the local economy but goes to overseas companies that own the hotels and resorts. Seventy per cent of Thailand's tourist earnings goes to foreign investors.

In some countries, such as Indonesia and Jamaica, local people employed in tourism earn about 1 per cent of the income of their guests. Resort workers in Fiji earn the equivalent of $2 an hour.

Although tourism can provide employment opportunities, there is also the potential for exploitation, with workers being poorly paid for long hours of work.

Tourism damages culture and community, and exploits locals.

Child prostitution aimed at foreign tourists is an ongoing problem in many countries such as Thailand and Cambodia.

Tourists often ignore cultural dress codes. This can lead to social tension in places such as Bali and many Polynesian countries.

Over-commercialisation creates artificial culture. Traditional dances are shortened to accommodate tourists; sacred objects and artefacts are mass produced for the tourism industry.

There is growth in crime rates as rich tourists interact with poor locals.

8.8 EXERCISES

Geographical skills key: GS1 Remembering and understanding **GS2** Describing and explaining **GS3** Comparing and contrasting **GS4** Classifying, organising, constructing **GS5** Examining, analysing, interpreting **GS6** Evaluating, predicting, proposing

8.8 Exercise 1: Check your understanding

1. **GS2** One criticism of tourism is that it is 'over-commercialised'. Explain your understanding of this term.
2. **GS2** Tourism provides both direct and indirect employment. Provide an example of each.
3. **GS2** Explain how tourism can improve the living conditions for individuals in developing nations.
4. **GS2** Explain how tourism may lead to an increase in the crime rate in a popular tourist destination.
5. **GS2** Explain how tourism can lead to the preservation and conservation of ancient ruins and the creation of nature reserves.

▶

8.8 Exercise 2: Apply your understanding

1. **GS4** Study **FIGURES 2** and **3**.
   (a) Create a table like the one shown below to classify the impacts of tourism as social, economic or *environmental*.

Impacts of tourism	Social/cultural	Economic	Environmental
Positive			
Negative			

   (b) Rank each of the impacts from most to least impactful. Write a paragraph to explain your rankings.
2. **GS2** Using the table you created in question 1, select a negative impact from each category. Explain the *scale* of each impact and devise a strategy for *sustainable* tourism.
3. **GS5 FIGURE 4** shows how the tourist dollar can flow from one job to the next. The jobs in the centre of the diagram interact directly with the tourist, while those on the outside do not.
   (a) Copy the diagram into your workbook at an enlarged size. Complete it by adding other jobs.
   (b) Study your completed diagram and write a paragraph explaining the *interconnection* between tourism and the economy.
   (c) Repeat this exercise looking at either the social or *environmental* impacts.

FIGURE 4 One view of tourism

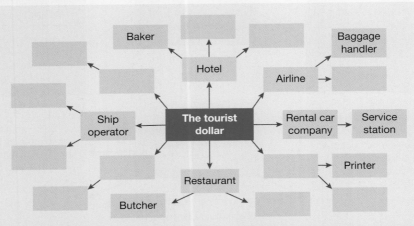

4. **GS6** The type of *interconnection* shown between industries in **FIGURE 4** is sometimes called the multiplier effect. Explain what you think this means.
5. **GS6** Which of the following would be the best to develop as a tourist resource in your region: art gallery, museum, cinema complex or sports stadium? Justify your answer.

Try these questions in learnON for instant, corrective feedback. Go to www.jacplus.com.au.

8.9 Managing environmental impacts — eco-friendly tourism

8.9.1 What is ecotourism?

Tourism has the capacity to benefit environments and cultures or destroy them. **Ecotourism** has developed in response to this issue. The aim is to manage tourism in a sustainable way. This might be through educational programs related to the environment or cultural heritage, or through controlling the types and locations of tourist activities or the number of tourists visiting an area. Ecotourism is the fastest growing sector in the tourism industry, increasing by about 10 to 15 per cent per year.

Ecotourism differs from traditional tourism in two main ways.

- It recognises that many tourists wish to learn about the natural environment (such as reefs, rainforests and deserts) and the cultural environment (such as indigenous communities).
- It aims to limit the impact of tourist facilities and visitors on the environment.

FIGURE 1 An ideal ecotourism resort

(A) The natural bush is retained and native plants are used to revegetate or landscape the area.

(B) Composting toilets treat human waste, and worm farms consume food waste. Water is treated with ultraviolet light rather than chlorine. Recycling is practised; for example, greywater is used in irrigation and toilet systems.

(C) Visitors are encouraged to improve and maintain the environment by using paths or planting trees.

(D) Buildings blend in with the natural landscape, and local materials are used. Buildings are often raised to prevent damage to plant roots. During construction, builders prevent contamination of the local environment by having workers change shoes and by washing down equipment to keep out foreign organisms.

(E) Local organically grown produce is used, and craft markets and stalls might also be established and run by indigenous communities, supporting the local economy, creating jobs and reducing poverty.

(F) There is no golf course, because of the water that would need to be used and the pesticides it would require.

(G) Low-impact, non-polluting transport such as bicycles is provided for guests.

(H) Walking trails include educational information boards.

(I) An information centre helps visitors understand the environment. Local indigenous people are employed to educate visitors about their culture.

(J) Electricity is generated through solar panels on the roofs of eco-cabins.

(K) Boardwalks are built over sensitive areas such as sand dunes to protect them from damage. Boardwalks might also be constructed in the tree canopies.

(L) Trained guides educate tourists about coral reefs and native vegetation, and show visitors how to minimise their impact.

One of the most famous examples of wildlife-based ecotourism in Australia is Monkey Mia in Western Australia. Here the wild dolphins come into shore and tourists are able to feed, swim with and touch them. **FIGURE 2** shows some of the regulations in place to manage this experience for the mutual benefit of tourists and the marine wildlife.

FIGURE 2 Regulations for contact with dolphins

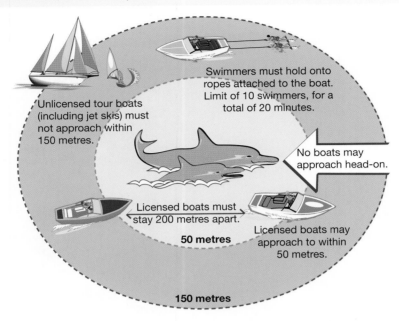

Unlicensed tour boats (including jet skis) must not approach within 150 metres.

Swimmers must hold onto ropes attached to the boat. Limit of 10 swimmers, for a total of 20 minutes.

No boats may approach head-on.

Licensed boats must stay 200 metres apart.

Licensed boats may approach to within 50 metres.

50 metres

150 metres

8.9.2 Are zoos and aquariums eco-friendly?

The development of zoos over time

A zoo is a place where animals are held in captivity and are put on display for people to view. The concept of zoos is not new; wall carvings provide evidence that ancient civilisations had zoos. Some of the earliest records date back to 2500 BCE when wealthy aristocrats and rulers in Egypt and Mesopotamia had their own private collections called menageries. They arranged expeditions to distant lands to bring back both terrestrial and marine creatures. They hired people to care for their collections and ensure that their animals not only thrived but also reproduced.

The first public zoo was built in 1793 in Paris, France. People not only wanted to be able to see exotic foreign animals but also study them for scientific purposes. However, these early zoos have been described as resembling a museum with living animals that were kept in small display areas rather than natural habitats (see **FIGURE 3**).

In recent years the design of zoos has undergone a major transformation in many parts of the world. People have become more aware of the plight of animals that are kept in captivity. The trend is towards giving animals more space and redeveloping enclosures to mimic their natural environment (see **FIGURE 4**). Education programs are also integrated to provide information about the threats and challenges that exist in the wild. Conservation strategies are highlighted, including what individuals may do to help. Captive breeding programs are also part of a much wider conservation strategy.

Today the definition of a zoo can be extended to include wildlife reserves, petting zoos, aquariums and aviaries, where care is taken to reproduce natural environments including cold habitats for animals such as polar bears and heated enclosures with regulated humidity for species from tropical areas.

The redevelopment of zoos and the focus on conservation rather than purely on human entertainment has meant that they are considered to have some environmental benefit. Although the International Union for Conservation of Nature (IUCN) Red List of endangered and threatened species continues to grow, zoos have been instrumental in ensuring that some species have been brought back from the brink of extinction.

FIGURE 3 Melbourne Zoo, 1963

FIGURE 4 Melbourne Zoo, 2016

Przewalski's horse

Przewalski's horse is a rare and endangered wild horse, native to the Gobi Desert in Mongolia. In 1969 it was declared extinct in the wild, with the only remaining populations being held in the Munich and Prague zoos. At one stage only 15 of the animals remained.

A captive breeding program was devised to not only increase numbers but also ensure the genetic viability of the offspring produced. The program, which continues today, had astounding success.

Today, there are about 1500 Przewalski's horses in captivity and 400 in the wild after they were successfully reintroduced into a protected reserve in Mongolia. Despite the harsh conditions of their natural environment and setbacks such as the brutally cold winter of 2009, which saw the wild population fall by half, they have continued to thrive.

Helmeted Honeyeater

Victoria's bird emblem, the Helmeted Honeyeater, is listed as critically endangered, with numbers falling to as low as 50 in the wild. There are currently only three small semi-wild populations, located to the east of Melbourne in the Yellingbo Conservation Reserve, which consists largely of swamp forest (see **FIGURE 5**). Healesville Wildlife Sanctuary (part of Zoos Victoria) has a captive breeding program that aims to increase the number of birds in the wild to ten interconnected and stable breeding colonies.

Due to the risk of fire, the sanctuary also has a management plan in place that includes relocating its inhabitants to Melbourne in the event of fire sweeping through the area. The Helmeted Honeyeaters held at Healesville are the only viable populations that could be used to repopulate the Yellingbo Conservation Reserve if it were devastated by fire.

FIGURE 5 Helmeted Honeyeater distribution map

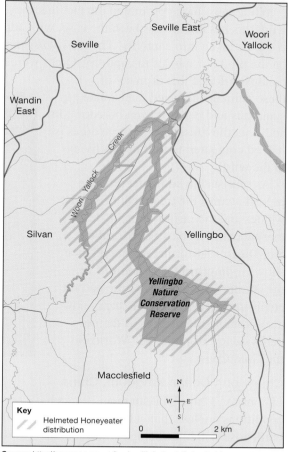

Source: http://www.zoo.org.au/healesville/animals/helmeted-honeyeater
The State of Victoria, Department of Environment, Land, Water & Planning 2016

  Resources

Google Earth: Yellingbo Conservation Reserve

Are all zoos eco-friendly?

While zoos primarily promote themselves as acting in the best interests of the wildlife they keep, sometimes this is not necessarily the case. The scenario outlined in **FIGURE 6** provides one example of a situation where the best interests of the animal were not taken into account.

FIGURE 6 The case of Marius the giraffe

Last weekend, a healthy juvenile male reticulated giraffe at the Copenhagen Zoo was killed. His name was Marius. The reason given was that his genes were already sufficiently represented in the giraffe population across the zoos of the European Association of Zoos and Aquariums (EAZA) — his brother lives in a zoo in England, for example — making him a so-called 'surplus animal.' Despite the international outcry against it, the giraffe was euthanized, a necropsy was performed by scientists while educators explained the dissection to the gathered crowd, and hunks of meat were fed to the zoo's lions, polar bears, and other carnivores.

Source: Jason G Goldman

Marine parks

Marine parks are similar to zoos except that they are home to marine creatures. They were a boom industry in the mid to late twentieth century, with many offering not just the opportunity to view marine animals but also to watch them perform.

While most parks promoted the fact that they only housed animals born in captivity or rescued from the wild, this has not always been the case, with adult orcas, for example, sometimes killed so that their young could be taken into captivity. The *Free Willy* movies, which first screened in the 1990s, focused attention on the plight of whales in captivity. Orcas were often housed in pools that were inadequate in size; the collapsed dorsal fins of many of these animals indicated inadequate standards and perhaps even boredom. In addition, the lifespan of captive orcas is halved compared to the species living in the wild.

As tourists boycotted facilities in protest against the treatment of killer whales and the visitor numbers fell, marine parks such as Sea World in San Diego, California, revamped their shows to improve the conditions of the animals and emphasise the natural environment. Some marine parks now also promote themselves as theme parks — for example, Sea World on the Gold Coast.

FIGURE 6 Sea World in California is home to orcas.

Chimelong Ocean Kingdom, China

In contrast, China's ocean theme parks have experienced a tourist boom despite the fact that their living conditions are less than adequate. It is thought that across China 872 cetaceans are held, including bottlenose dolphins, beluga whales, orcas and sea lions, all taken directly from the wild and sold on the black market at prices ranging from US$50 000 to US$1 000 000.

Chimelong Ocean Kingdom is home to whales, dolphins, polar bears, Arctic foxes, walruses and seals held in tanks and enclosures described as inadequate in terms of meeting their social and behavioural needs. Their diet is similarly inadequate and animal conservationists report that cramped living conditions are resulting in aggressive behaviours rarely seen in the wild.

In their natural Arctic and sub-Arctic environment, beluga whales are social animals that live in pods varying in size from a few to a couple of hundred individuals. They may travel up to 160 kilometres in a day and dive to depths of 300 metres. In captivity, they are confined to shallow tanks that only allow them to aimlessly circle their enclosure and are expected to perform tricks that are not a part of their natural behaviour.

Explore more with myWorldAtlas

Deepen your understanding of this topic with related case studies and questions.
- Investigate additional topics > Tourism > **Kakadu National Park**
- Investigate additional topics > Managing environments > **Wilsons Promontory**
- Investigate additional topics > Tourism > **Ningaloo Reef and ecotourism**

on Resources

Interactivity Sustainable sightseeing (int-3336)
Weblink Zoos Victoria

8.9 INQUIRY ACTIVITIES

1. Visitors to ecotourism resorts are often attracted by brochures that emphasise the resort's **environmental** policies. These brochures also set out guidelines to follow in order to minimise visitor impact.
   (a) Design and produce a brochure for the ecotourism resort illustrated in **FIGURE 1**. Use ICT tools and techniques to maximise the brochure's impact.
   (b) Add another eco-activity to the island and devise strategies to educate tourists and minimise their impact on the **environment**. **Classifying, organising, constructing**
2. Use the **Zoos Victoria** weblink in the Resources tab.
   (a) Investigate what is being done to fight extinction.
   (b) Investigate one of the threatened species featured on the website.
   (c) Prepare an annotated visual display of your findings. **Examining, analysing, interpreting** **Classifying, organising, constructing**
3. Select and investigate one of the species mentioned in this subtopic. Write a report that compares the species' natural habitat existence with that of being held in captivity. **Examining, analysing, interpreting**

8.9 EXERCISES

Geographical skills key: GS1 Remembering and understanding **GS2** Describing and explaining **GS3** Comparing and contrasting **GS4** Classifying, organising, constructing **GS5** Examining, analysing, interpreting **GS6** Evaluating, predicting, proposing

8.9 Exercise 1: Check your understanding

1. **GS1** How does an ecotourism resort differ from a traditional tourist resort?
2. **GS2** Consider the ecotourism example of Monkey Mia, Western Australia.
   (a) What rules and other techniques are used to control the interaction between dolphins and tourists?
   (b) Predict potential problems that might occur between dolphins and tourists.
   (c) Do you think this is an example of **sustainable** ecotourism? Give reasons for your answer.
3. **GS1** What is a zoo?
4. **GS1** Outline the different types of zoos and the positive and negative impacts they may have on wildlife.
5. **GS2** Explain what you understand by the term *captive breeding program*.

8.10 Cultural tourism

8.10.1 Defining cultural tourism

According to the UN World Tourism Organization,

> Cultural tourism is a type of tourism activity in which the visitor's essential motivation is to learn, discover, experience and consume the tangible and intangible cultural attractions/products in a tourism destination.
>
> These attractions/products relate to a set of distinctive material, intellectual, spiritual and emotional features of a society that encompasses arts and architecture, historical and cultural heritage, culinary heritage, literature, music, creative industries and the living cultures with their lifestyles, value systems, beliefs and traditions.

Cultural tourism is not a new thing — it has long been a factor in many people's reasons for travel. Visits to places like Port Arthur in Tasmania, Sovereign Hill in Ballarat, the Colosseum in Rome or the Pyramids in Egypt can be considered cultural tourism, as people endeavour to learn about and connect with the past. This type of tourism prompts us to preserve and protect our heritage.

Visiting art galleries, such as the Louvre in Paris, attending music and theatre performances, or even undertaking a cooking course in another place are also examples of cultural tourism activities that broaden our knowledge and understanding of our world and its people.

In 2017, more than two-fifths of international visitors to Australia included a cultural event in their itinerary and a third participated in a heritage event. Cultural tourists also tend to stay longer. In 2017, international visitors to Victoria who attended a cultural event stayed for 25 nights, compared with 23 nights for those who did not.

Globally, cultural tourism is on the rise as people return home, undertake a pilgrimage, or simply want to experience a significant cultural event in another place. Examples include the following.

- The Day of the Dead — originating in Mexico, the festival celebrates the dead, who have awakened to celebrate with loved ones before continuing their spiritual journey (see **FIGURE 1**). The festivities span three days, and the public holiday encourages people to remember and pray for family and friends who have passed away. The custom has now spread to other places such as the United States and other countries in Latin America.
- The ancient religious festival of Holi, marking the arrival of spring, celebrates the start of a plentiful spring harvest. Originating in the predominantly Hindu nations of India and Nepal, it is also referred to as the Festival of Colours, because of the traditional practice of throwing colours at the Emperor. It has now also spread to other parts of Asia, the Caribbean, North America and South Africa.

- The Hajj pilgrimage to the sacred city of Mecca, located in Saudi Arabia, is a practice dating back to the ancient prophets. With the expectation that it will be made at least once in every Muslim person's lifetime, the Hajj is an enormous gathering, attracting 3 to 5 million people each year (see **FIGURE 2**). It occurs over 5 or 6 days in the last month of the Islamic calendar.
- In many cultures where Christianity is the predominant religion, people come together to celebrate Christmas, commemorating the birth of Christ, and Easter, to remember his resurrection.

Whatever the reason, the mass movement of people associated with these events has a significant impact on both people and places.

FIGURE 1 Thousands gather to attend the Day of the Dead parade in Mexico City each year.

FIGURE 2 Muslims from all over the world make the Hajj pilgrimage to Mecca, Saudi Arabia.

8.10.2 Thanksgiving

Thanksgiving is held each year in the United States on the fourth Thursday in November. It dates back to the seventeenth-century celebration of the harvest. Today it is a time for families to get together and give thanks for what they have.

The Thanksgiving holiday period runs from Wednesday to Sunday. As millions of people travel across the United States, transport systems are stretched to their limits, creating traffic congestion and delays. Because the holiday season is so close to the start of winter, the weather can further complicate people's travel plans, especially for those who live in the colder northern states. Early winter storms can bring ice and snow, resulting in airport closures and impassable roads.

The average American will spend around 21 per cent of their Thanksgiving budget on travelling to their destination, whether by car, air or other means (see **FIGURE 3**).

FIGURE 3 Modes of transport used for Thanksgiving travel

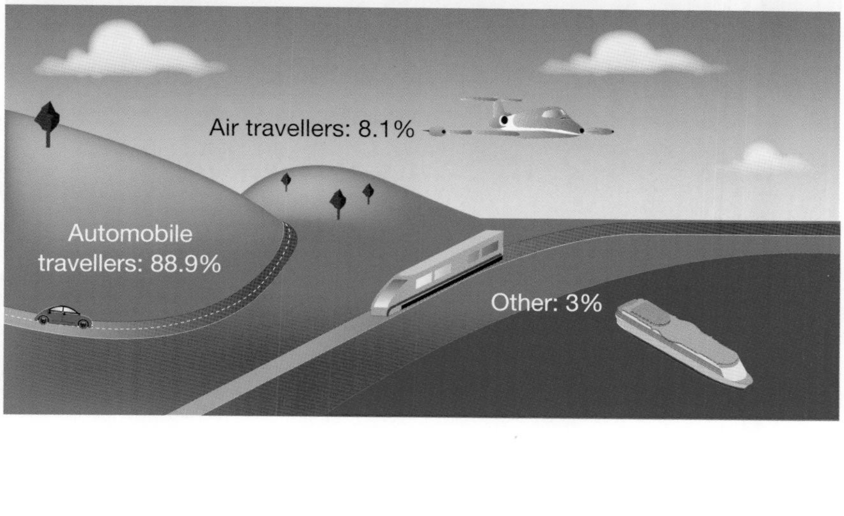

Air travellers: 8.1%

Automobile travellers: 88.9%

Other: 3%

8.10.3 Chinese New Year

Chinese New Year is the longest and most important of the traditional Chinese holidays. Dating back centuries, it is steeped in ancient myths and traditions. The festivities begin on the first day of the first month in the traditional Chinese calendar, and last for 15 days. They conclude with the lantern festival on Chinese New Year's Eve, a day when families gather for their annual reunion dinner. It is considered a major holiday, and it influences not only China's geographical neighbours but also the nations with whom China has economic ties.

The date on which Chinese New Year occurs varies from year to year. This date coincides with the second **new moon** after the Chinese **winter solstice**, which can occur any time between 21 January and 20 February.

Chinese New Year, or Lunar New Year, is celebrated as a public holiday in many countries with large Chinese populations or with calendars based on the Chinese lunar calendar (see **FIGURE 4**). The changing nature of this holiday has meant that many governments have to shift working days to accommodate this event.

In China itself, many manufacturing centres close down for the 15-day period, allowing tens of millions of people to travel from the industrial cities where they work to their hometowns and rural communities. This means that retailers and manufacturers in overseas countries such as the United States and Australia have to adjust their production and shipping schedules to ensure they have enough stock on hand to deal with the closure of factories in China. For those shopping online, delays in delivery are to be expected during this period.

FIGURE 4 To ensure prosperity and good fortune in the year ahead, parades, dragons and lion dances feature in Chinese New Year celebrations.

The logistics of moving millions

Chinese New Year has been described as the biggest annual movement of people in China. Over a five-day period, an average of 80 million journeys are recorded in the last-minute dash to make it home for the traditional family celebrations – a total of 400 million people on the move in just five days!

Although incomes have risen for middle-class citizens in China, most people elect to travel by road as they do not want to stand in long queues for hours or even days to purchase bus or rail tickets. In 2018, over the 40-day Spring Festival period that encompasses Chinese New Year, 2.4 billion trips were made by road, 389 million trips were made by train and 65 million trips were made by plane. Airlines scheduled an additional 200 flights to cope with the demand. It is not uncommon for commuters to add hundreds, or even thousands of kilometres to their journey; one airline passenger flew from Beijing to Kunming in Southern China via Bangkok in Thailand because there were no direct flights. Weather conditions and the impact of additional flights competing for the same amount of air space make delays inevitable.

Weather conditions can also impede rail and car travel. In 2017 a cold snap saw highways in central China covered in ice; this was further complicated by heavy fog making road travel close to impossible. In 2016, almost 100 000 people were left stranded at railway stations after ice and snow in other parts of the country caused long delays. Fifty-five trains in Shanghai and 24 in Guangzhou were unable to leave their respective stations when China was struck by a record-breaking cold snap (see **FIGURE 5**). Almost 4000 police and security guards were called in to keep order.

Late in 2018, ten new railways were added to the rail network to expand the length of China's high-speed railway — the second-largest in the world behind the United States. At its peak, the online rail booking system had to cope with 1000 bookings per minute! High demand also leads to high prices, and scalpers were quick to cash in, charging double or even triple the usual ticket cost.

FIGURE 5 Travel chaos as crowds swell outside Guangzhou station after bad weather causes long delays

For many, motorbike travel is the cheapest way to return home, with some making journeys in excess of 400 kilometres. Motorbikes offer not only a cost saving, but also a time saving. Although China boasts one of the world's largest road networks, with almost 98 000 kilometres of motorway, when 2.4 billion people take to the roads, congestion is inevitable. To ensure safety and improve traffic flow, 170 000 additional police in an extra 60 000 police vehicles are mobilised.

With such challenges to moving around China during this period, it is no wonder that a growing trend favoured by more than 7 million Chinese is to celebrate the New Year by travelling abroad to over 90 countries. Others are now electing not to travel at all, instead choosing to work through the holiday period to take advantage of increased pay rates on offer. In response to an increasing trend in takeaway food orders during the festivities, some employers in the hospitality industry are offering delivery drivers triple pay to work on Chinese New Year.

on Resources

🔗 **Weblink** Thanksgiving

8.10 INQUIRY ACTIVITY

a. As a class, brainstorm a list of cultural or celebratory events that occur in Australia.
b. Use the internet to find out more about either Chinese New Year or Thanksgiving. Investigate the history, myths and traditions associated with your chosen event. Prepare an annotated visual display comparing your findings with a cultural or celebratory event in Australia. Make sure you include references to the *scale* of your chosen event and the *place* in which it occurs. **Examining, analysing, interpreting Classifying, organising, constructing**

8.10 EXERCISES

Geographical skills key: GS1 Remembering and understanding **GS2** Describing and explaining **GS3** Comparing and contrasting **GS4** Classifying, organising, constructing **GS5** Examining, analysing, interpreting **GS6** Evaluating, predicting, proposing

8.10 Exercise 1: Check your understanding

1. **GS2** In your own words, explain what is meant by the term *cultural tourism*.
2. **GS2** Why are Thanksgiving and Chinese New Year regarded as cultural events?
3. **GS2** Explain why Chinese New Year leads to industries shutting down for 15 days.
4. **GS2** What is a *pilgrimage*?
5. **GS2** Describe the impact the weather might have on a cultural event.

8.10 Exercise 2: Apply your understanding

1. **GS5** Write a paragraph explaining how cultural events can *change* people, *places* and the *environment*.
2. **GS3** Answer the following to reflect on Thanksgiving and Chinese New Year travel.
   (a) What is the preferred mode of transport for Thanksgiving and for Chinese New Year? Suggest reasons for differences in travel arrangements. In your response, include reference to the *scale* of movement.
   (b) Make a list of problems associated with the mass movement of people.
   (c) Select one of the problems you have identified and explain the impact it might have on people, *places* and the *environment*. Suggest a strategy for the *sustainable* management of this problem in order to reduce its impact.
3. **GS5** Write a paragraph describing a traditional cultural event that you and your family celebrate. Is it an example of cultural tourism? Give reasons for your answer.
4. **GS6** Explain the impact Chinese New Year might have on a clothing import business in Australia. In your answer, explain what a business owner might need to do to ensure their business is not affected by this event.
5. **GS5** Some cultural events, such as Thanksgiving, occur at approximately the same time each year, whereas others such as Chinese New Year vary more in their timeframe. Explain why this is so.

Try these questions in learnON for instant, corrective feedback. Go to www.jacplus.com.au.

8.11 Tourism and sport

8.11.1 How are tourism and sport connected?

Sport tourism involves people travelling to view or participate in a sporting event or sporting pursuits. Tourism in which someone travels to either actively participate in or watch a competitive sport as the main reason for their travel is known as **hard sport tourism**. Tourism in which someone participates in recreational and leisure activities, such as skiing, fishing and hiking as part of their travel is known as **soft sport tourism**. A common trait in all sports tourists is their passion for the sport and a willingness to spend money to indulge this passion.

FIGURE 1 On the trail — soft sport tourism

Sport tourism is an expanding sector of the tourism industry, estimated to add $800 billion to global economies each year. It is estimated that between 12 million and 15 million international trips are made to view sporting events. But what impact does this have on people and places?

Governments spend millions of dollars to attract people to sporting events such as the Olympics, the cricket, the FIFA World Cup and motor racing events, to name just a few. These events also trigger:

- construction of new stadiums
- expansion and upgrades of transport networks
- improvements to airport facilities
- clean-ups of cities in readiness for the arrival of tourists.

8.11.2 Are the Olympics a tourist bonanza?

Major sporting events such as the Olympic Games translate into improved infrastructure, and provide the host city with considerable international exposure, but this comes at a substantial cost (see **FIGURE 2**). Does this bring in more tourists and justify the capital outlay?

FIGURE 2 Olympics expenditure (summer and winter), including infrastructure

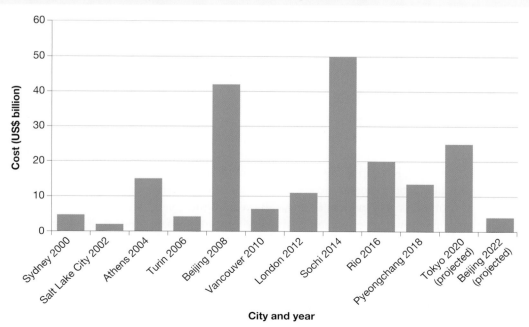

The general consensus among economists is that the costs associated with hosting a one-off major event, such as the FIFA World Cup or the Olympic Games, generally exceed the value of any anticipated long-term benefits. The 2016 Rio Olympic Games were plagued by political and economic controversy. As the cost of hosting the Games blew out to $20 billion, Brazil was plunged into recession. Issues such as the Zika virus epidemic, a Russian doping scandal and high levels of pollution also threatened to derail the games. On the plus side, upgrades to the public transport network and sewerage system, cleaning up the pollution at Guantanamo Bay, and the construction of nine new permanent venues and seven temporary venues delivered a boost to the construction industry. However, now that the Games are long past, Rio has a surplus of venues that it no longer needs. Plans to sell them off failed due to a lack of buyer interest, resulting in many venues, such as the Olympic pool and Maracanã Stadium (site of the opening and closing ceremonies) falling into a state of disrepair. The athletes' village, which housed 10 000 athletes in 3604 apartments, is largely empty, with only 7 per cent of the apartments sold.

In the United Kingdom during the 2012 Olympics year, statistics for August showed 5 per cent *fewer* visitors than in the previous year. Tourism spending, however, went up by 9 per cent, in part because of spending on Olympics tickets. In addition, many UK residents chose to holiday overseas rather than remain at home during the Olympic Games. Organisers were also frustrated by the number of empty seats in many of the venues. On the plus side, however, building the Olympic village provided a £6 billion boost to the building and construction industry.

FIGURE 3 The opening ceremony of the London Olympic Games

But what happens to the people who originally lived on the site of the new venues and athletes' village? Quite simply, they are moved on. While they may receive some compensation, land values go up in the shadow of renewed development. Residents simply cannot afford to live in the new developments, nor can they afford to renovate their existing dwellings. In the lead-up to the Beijing Olympics, 1.5 million Chinese people were forced out of their homes to make way for Olympics venues.

Once the event is over, many of the stadiums are underused, and it can take years to recover from the cost of staging the event. For instance, the city of Montreal in Canada, which hosted the games in 1976, took 30 years to pay back the equivalent of US$6 billion (in today's money) in Olympic spending.

8.11.3 Other sports events

It has generally been accepted that regular sporting events can have financial benefits for the host location. Many international tourists visiting the United Kingdom, for instance, include a sporting event on their itinerary. Most popular is soccer, because of the opportunity to see some of the world's most talented athletes playing in some the UK's top teams. Overall, sports tourists stay longer and are not deterred by the weather. Sporting fans also tend to spend more than the rest of the tourist population.

The popularity of football is also evident in Australia, where three separate codes (AFL, soccer and rugby league) attract huge crowds every week, and many fans are prepared to travel interstate to watch their teams play.

But it is not just football that attracts the crowds. The English cricket team, for example, is followed around the world by its unofficial cheer squad — the Barmy Army. Many Australian fans participate in a range of organised sporting tours each year, taking in some of the biggest events both at home and abroad involving, in addition to cricket, sports such as tennis, rugby and golf.

FIGURE 4 The Barmy Army are English cricket fans who travel the world to cheer on the English cricket team.

Explore more with my**World**Atlas

Deepen your understanding of this topic with related case studies and questions.
• Investigate additional topics > Australia's links with the world > **Sport**
• Investigating Australian Curriculum topics > Year 9: Geographies of interconnections > **The FIFA World Cup**

 Resources

Interactivity Are the Olympic Games worth gold? (int-3337)

8.11 INQUIRY ACTIVITY

Phillip Island is located 100 kilometres south-east of Melbourne, Victoria, and is linked to the mainland by a bridge. The area is popular for its beaches and wildlife, but it is also home to a Grand Prix racing circuit that stages a variety of motor sports throughout the year. Collectively, more than $110 million is generated annually from the circuit's car and bike activities. Three events — the Moto GP, V8 Supercars and Superbikes — bring in around $80 million. Each of these events brings more than 65 000 people to the island.

a. What facilities are needed to cater for such a large influx of people?
b. **FIGURE 4** in subtopic 8.8 shows how the tourist dollar can flow from one job to the next. Complete a diagram like this for the Phillip Island Grand Prix circuit.
c. With a partner, brainstorm a list of negative consequences that might result from having a Grand Prix circuit on Phillip Island. Make sure you consider the impact on people and the *environment*, as well as the *scale* of such effects.
d. Write a paragraph explaining the *interconnection* between the location of sporting facilities and their impact on people and *places*.
e. Do you think this is an example of *sustainable* tourism? Justify your point of view.

Examining, analysing, interpreting

8.11 EXERCISES

Geographical skills key: GS1 Remembering and understanding **GS2** Describing and explaining **GS3** Comparing and contrasting **GS4** Classifying, organising, constructing **GS5** Examining, analysing, interpreting **GS6** Evaluating, predicting, proposing

8.11 Exercise 1: Check your understanding

1. **GS1** Define *hard sport tourism*.
2. **GS1** Define *soft sport tourism*.
3. **GS2** Consider the concept of sport tourism.
   (a) Is someone who goes to a local football match a sport tourist? Explain.
   (b) What if that person travels interstate? Explain.
4. **GS2** Brainstorm a list of sports on which money might be spent to attract tourists. Categorise these as either hard or soft sport tourism events.
5. **GS4** Compile a table that highlights the positives and negatives of sport tourism.
6. **GS2** From the table you created in question 5, choose two positives and two negatives. For each, explain the impact it has on people and *places*.

8.11 Exercise 2: Apply your understanding

1. **GS6** Suggest a reason why some of the stadiums built for the Rio Olympic Games were temporary rather than permanent.
2. **GS1** Explain how hosting a major international sporting event can lead to improvements in infrastructure.
3. **GS6** Identify a financial cost and a financial benefit of hosting a major sporting event. In your opinion does the benefit outweigh the cost? Give reasons for your answer.
4. **GS6** Queensland has recently expressed interest in hosting the 2032 Olympic Games. Considering the pros and cons of hosting such an event, what advice would you give to the Queensland government?
5. **GS6** A leading economist recently said, 'Major events such as the Olympics should be hosted by developed countries; the cost to developing nations is too great.' To what extent do you agree with this statement? Explain your view.

Try these questions in learnON for instant, corrective feedback. Go to www.jacplus.com.au.

8.12 Thinking Big research project: Design a 7-day cruise adventure

online only

SCENARIO

Mystic Cruises is about to add a new cruise ship to its fleet. As part of the company's cruise development team, you need to design a 7-day cruise, including exotic ports of call and shore excursions that allow cruise guests to take in the sites and culture of the places they visit.

Select your learnON format to access:

- the full project scenario
- details of the project task
- resources to guide your project work
- an assessment rubric

 Resources

ProjectsPLUS Thinking Big research project: Design a 7-day cruise adventure (pro-0194)

8.13 Review

8.13.1 Key knowledge summary
Use this dot point summary to review the content covered in this topic.

8.13.2 Reflection
Reflect on your learning using the activities and resources provided.

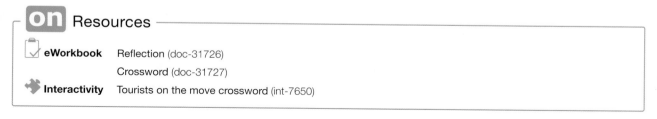

on Resources

eWorkbook	Reflection (doc-31726)
	Crossword (doc-31727)
Interactivity	Tourists on the move crossword (int-7650)

KEY TERMS

developed describes countries with a highly developed industrial sector, a high standard of living, and a large proportion of people living in urban areas

ecotourism tourism that interprets the natural and cultural environment for visitors, and manages the environment in a way that is ecologically sustainable

gross domestic product (GDP) the value of all the goods and services produced within a country in a given period of time (usually a year). It is often used as an indicator of a country's wealth.

hard sport tourism tourism in which someone travels to either actively participate in or watch a competitive sport as the main reason for their travel

infrastructure the facilities, services and installations needed for a society to function, such as transportation and communications systems, water pipes and power lines

Masai an ethnic group of semi-nomadic people living in Kenya and Tanzania

mature-aged describes individuals aged over 55

national park a park or reserve set aside for conservation purposes

new moon the phase of the moon when it is closest to the sun and is not normally visible

soft sport tourism tourism in which someone participates in recreational and leisure activities, such as skiing, fishing and hiking as part of their travel

winter solstice the shortest day of the year, when the sun reaches its lowest point in relation to the equator

9 Trade — a driving force for interconnection

9.1 Overview

Buy, swap, sell, give. Is the trade that occurs between different countries just a way of getting things?

9.1.1 Introduction

Trade, in the form of buying, swapping, selling and giving of goods and services, is a driving force that interconnects people and places all over the world. Trade has been going on since the beginning of human society. In contrast, international aid is a modern phenomenon, although countries have always had internal programs to help those in need. Trade and aid can bring people together to share the Earth's resources, but there can be problems when those resources are limited and, potentially, negative consequences for the environment. The big question is how to organise trade and aid so that they foster social justice and are fair and sustainable.

 **Resources**

☑ **eWorkbook**   Customisable worksheets for this topic

🎞 **Video eLesson** Trading places (eles-1724)

LEARNING SEQUENCE

9.1 Overview
9.2 How does trade connect us?
9.3 Australia's global trade
9.4 Food trade around the world
9.5 **SkillBuilder:** Constructing multiple line and cumulative line graphs `online only`
9.6 Impacts of globalisation
9.7 Making trade fair
9.8 Global connections through Australian Aid
9.9 The troubling illegal wildlife trade
9.10 **SkillBuilder:** Constructing and describing a flow map `online only`
9.11 **Thinking Big research project:** World Trade Fair infographic `online only`
9.12 **Review** `online only`

To access a pre-test and starter questions and receive immediate, **corrective feedback** and **sample responses** to every question, select your learnON format at www.jacplus.com.au.

9.2 How does trade connect us?

9.2.1 Trade in goods and services

The Earth's resources are not distributed evenly over space. For instance, some places may have an abundance of iron ore and others may have none. To solve this problem, nations have developed trade, allowing producers and consumers to exchange goods and services.

The system of trade has been around for a long time. Its earliest form was as barter at local markets or fairs. Merchants also used land and sea routes to access markets in foreign lands, where they exchanged goods for payment. More recently, air transport has become a means of trade, and the internet has made it possible to instantly exchange information. Today, we have a highly sophisticated, large-scale, global system of trade.

A modern example of the interconnection of trade is the production of the Airbus A380. To construct this plane, component parts must be purchased from different countries and transported over land and sea to reach their final assembly place in Toulouse, France (see **FIGURE 1**).

Goods and services, of which there are many, are generated by either processing Earth's resources (goods) or people doing things for each other (services). A good can be an item as simple as a loaf of bread or it can be as complex as a motor car. A service is not something you can hold in your hand; examples of a service are education in a school or the advice a doctor gives a patient. What types of goods and services do you use to support your lifestyle?

As seen in **FIGURE 2**, the processing of a resource into more complex goods can be a series of transitions, in which there is **value adding** at each level of industry (that is, its value increases). An important consideration in the production of goods and services is the impact on the environment.

FIGURE 1 The component parts routes of the Airbus A380

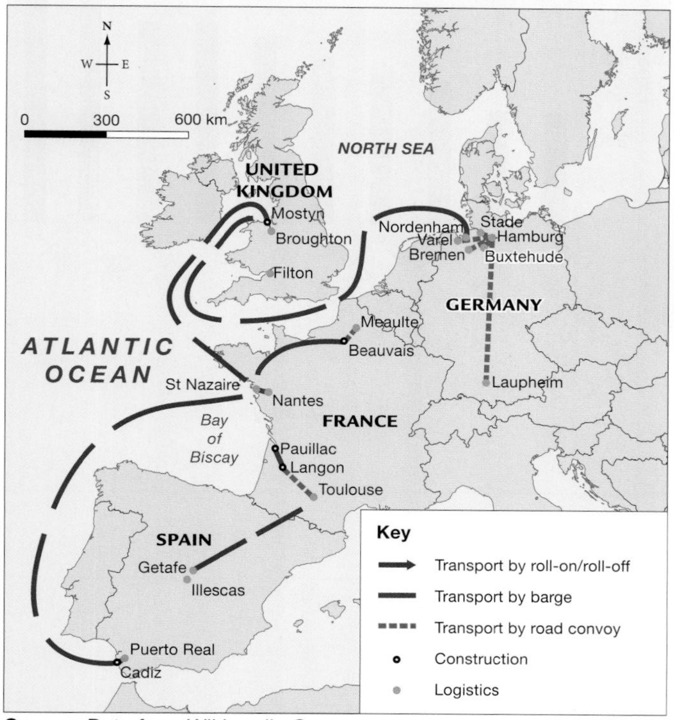

Source: Data from Wikimedia Commons

FIGURE 2 Four levels of industry

Primary industry
Takes natural resources from the Earth or grows them
Example: a farmer grows corn that is then transported to a canning factory.

Secondary industry
Makes products from natural resources
Example: a factory makes tins of corn and sells them to supermarkets.

Tertiary industry
Sells products or services
Example: a supermarket sells tins of corn and other products to consumers.

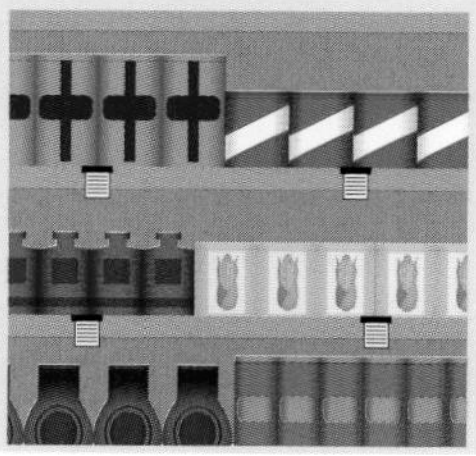

Quaternary industry
Sells knowledge and information
Example: a marketing analyst works out how best to position products, and sells this information to supermarkets.

9.2.2 How are goods and services consumed?

Household final consumption per person

If we tally the value or money spent on all goods and services such as food, cars, washing machines, electricity, water and gas, education, medical service expenses and entertainment within a country for a year, then divide this figure by the total population of the country, we obtain what is referred to as the household final consumption per person. This per-person dollar value can provide a general indication of the economic development and prosperity of a country.

The greatest consumers of goods and services on a per-person basis tend to be wealthy, industrialised countries, as shown in **FIGURE 3**. However, countries such as China and India also consume high levels of goods and services because they have very large populations. As would be expected, countries that are high-level consumers can have a significant impact on the environment, particularly in terms of energy use and waste production.

At the lower end of the scale of household final consumption per person, people in countries such as Niger and the Democratic Republic of the Congo spend $269 and $275 respectively per year, or around $0.75 per day per person (see **FIGURE 4**). This expenditure is mainly for food.

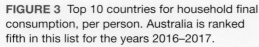

FIGURE 3 Top 10 countries for household final consumption, per person. Australia is ranked fifth in this list for the years 2016–2017.

FIGURE 4 Household final consumption expenditure per person for the 10 lowest ranked countries in the world for 2016–2017

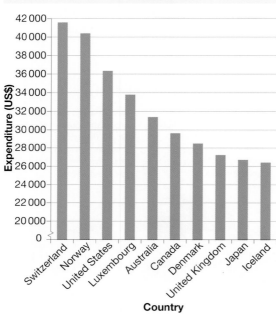

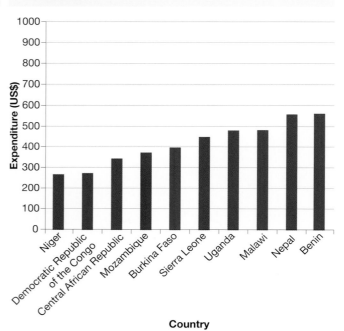

Resources

Interactivity Up, up and away! (int-3339)

9.2 EXERCISES

Geographical skills key: GS1 Remembering and understanding **GS2** Describing and explaining **GS3** Comparing and contrasting **GS4** Classifying, organising, constructing **GS5** Examining, analysing, interpreting **GS6** Evaluating, predicting, proposing

9.2 Exercise 1: Check your understanding

1. **GS2** What reasons can you suggest as to why goods and services are traded?
2. **GS1** Name the four levels of industry and give an example of a good as it moves through the production process.
3. **GS2** What reasons can you suggest for component parts of the Airbus A380 having to come from different *places* (countries)?
4. **GS2** Explain what is meant by the term *value adding*, as a product moves through the four levels of industry. Choose a product such as wheat or timber to explain this process.
5. **GS2** What are some of the impacts of a high level of consumption?

9.2 Exercise 2: Apply your understanding

1. **GS6** Suggest why the United States is one of the largest consumers of goods and services in the world.
2. **GS6** It has been claimed that countries such as China and India, with growing middle classes that are now eager for goods and services, will put a strain on world resources. How might a growing demand for energy sources in these countries affect the *environment*?
3. **GS6** How might a *change*, such as growth in Australia's population from 25 million to 40 million, affect Australia's trade?
4. **GS5** In 2016–17, Hong Kong, which is a Special Administrative Region (SAR) of China, was ranked eleventh in household final consumption per person, while the wider country of China was ranked eighty-fifth. Why might this be the case?
5. **GS5** What reasons can you give for people being able to survive on less than $300 per person per year in countries such as Niger and the Democratic Republic of the Congo?

Try these questions in learnON for instant, corrective feedback. Go to www.jacplus.com.au.

9.3 Australia's global trade

9.3.1 The coordination of trade

Australia is one of the 164 members of the World Trade Organization (WTO), which covers 95 per cent of global trade. The organisation promotes free and fair trade between countries and, since 2001, its Doha Development Agenda has aimed to help the world's poor by slashing **trade barriers** such as tariffs, quotas and farm subsidies.

The Department of Foreign Affairs and Trade (DFAT) coordinates trade agreements on behalf of the Australian government, and the Australian Trade Commission (Austrade) promotes the export of goods and services. In 2017–18, $587 billion or 73.5 per cent of Australia's total trade was with the member countries of the Asia-Pacific Economic Cooperation (APEC) forum.

FIGURE 1 People involved with trade

9.3.2 Australia's trading partners

China, Japan and the United States were Australia's top three two-way **trading partners** in 2017–18, accounting for nearly 43 per cent of total trade. **FIGURE 1** shows the value of imports and exports traded between Australia and its top ten trading partners. **TABLE 1** shows the total two-way trade value (imports and exports added together) of all goods and services traded with these ten countries.

FIGURE 2 Australia's top ten two-way trading partners 2017–18, value of imports and exports (A$ million)

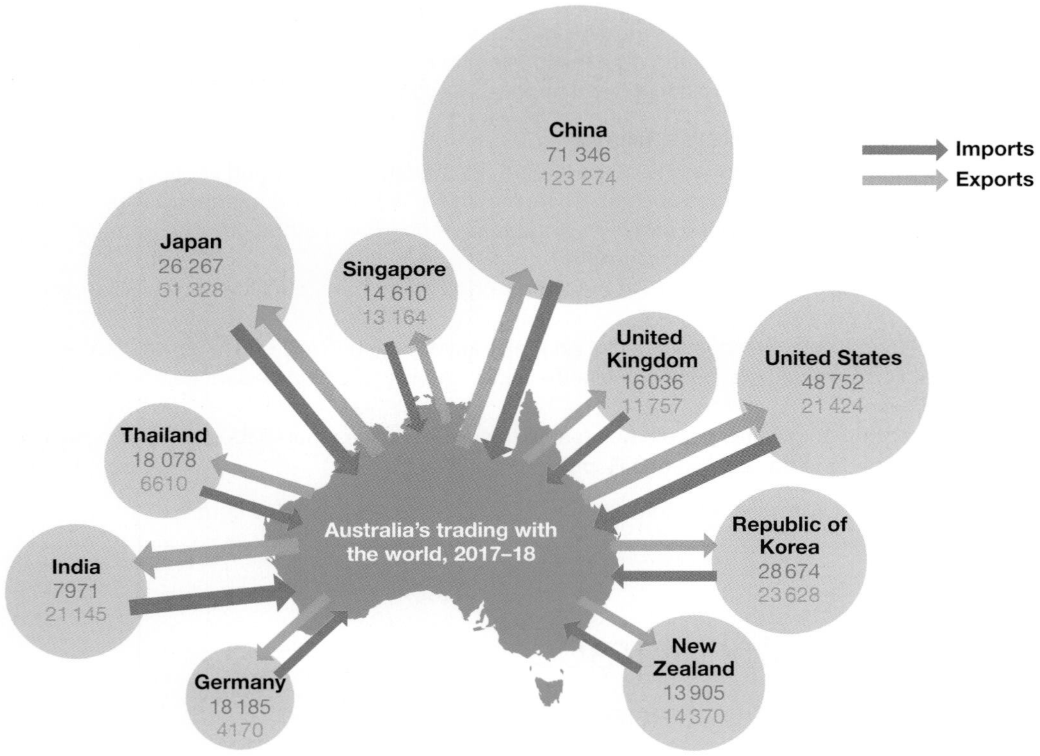

China
71 346
123 274

Imports
Exports

Japan
26 267
51 328

Singapore
14 610
13 164

United Kingdom
16 036
11 757

United States
48 752
21 424

Thailand
18 078
6610

Australia's trading with the world, 2017–18

Republic of Korea
28 674
23 628

India
7971
21 145

New Zealand
13 905
14 370

Germany
18 185
4170

TABLE 1 Australia's top 10 two-way trading partners, 2017–18, total two-way trade value (A$ million)

Rank		Goods	Services	Total	% share
	Total two-way trade	617 565	181 076	798 641	
1	China	174 451	20 169	194 620	24.4
2	Japan	71 348	6247	77 595	9.7
3	United States	44 018	26 159	70 177	8.8
4	Republic of Korea	49 300	3003	52 303	6.5
5	India	21 868	7248	29 116	3.6
6	New Zealand	17 282	10 993	28 275	3.5
7	United Kingdom	13 772	14 021	27 793	3.5
8	Singapore	17 308	10 465	27 773	3.5
9	Thailand	20 355	4333	24 688	3.1
10	Germany	17 230	5125	22 355	2.8

9.3.3 Australia's types of trade

Exports

Australia's export trade in 2017–18 was valued at $403 billion, and was dominated by the mineral products of iron ore and coal. Education-related and personal travel were Australia's leading services exports. See **FIGURE 3** for details of leading exports.

FIGURE 3 Australia's top ten exports of goods and services 2017–18 — value and share of total exports by sector

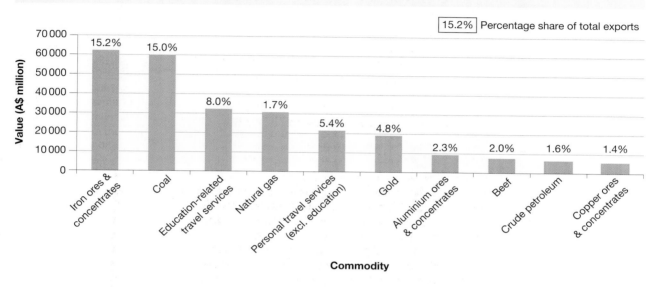

International students

A more recent high-level earner for Australia (now ranked as our third-highest export), is the category of 'education-related travel services', which for 2017–18 was valued at over $32 billion. In effect, education is a service export, in that students are paying for knowledge that they will take back to their home country.

Numbers of international students have grown significantly in recent years. In 2014, Australia hosted 450 000 international students; by 2018, there were 693 750 students from more than 200 countries studying in Australia (see **FIGURE 4**), making education a very important factor in our economy.

FIGURE 4 International students studying in Australia, 2018

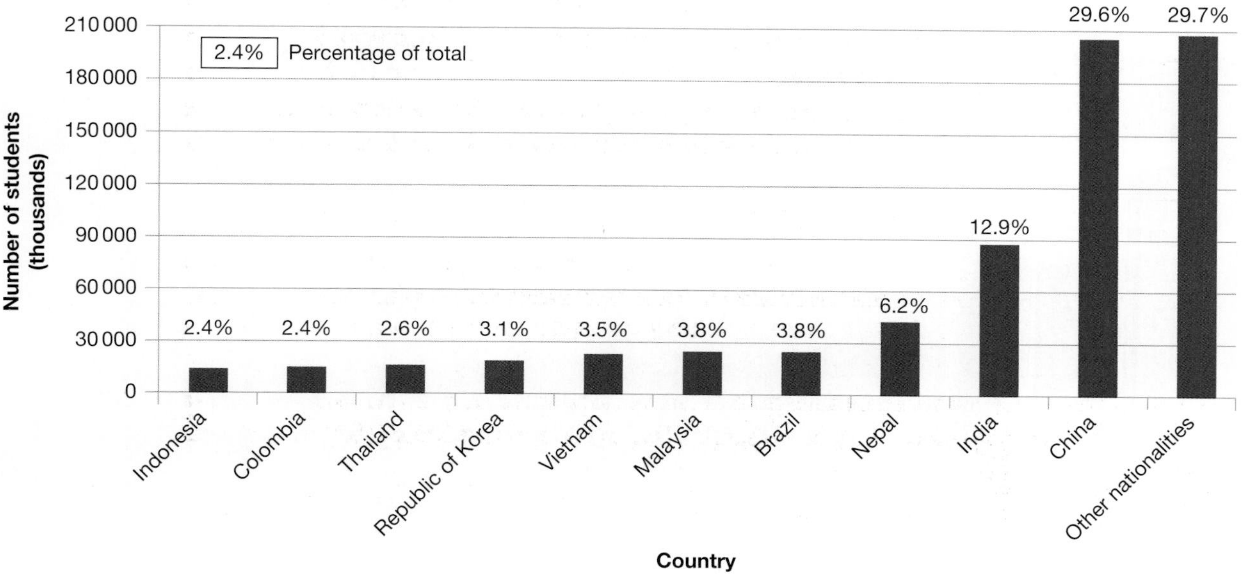

Imports

Like many countries, Australia is not self-sufficient in all goods and services. In 2017–18 Australia imported goods and services valued at over $395 billion. **FIGURE 5** shows the top ten commodities of this trade.

FIGURE 5 Australia's top ten imports of goods and services, 2017–18 (A$ million)

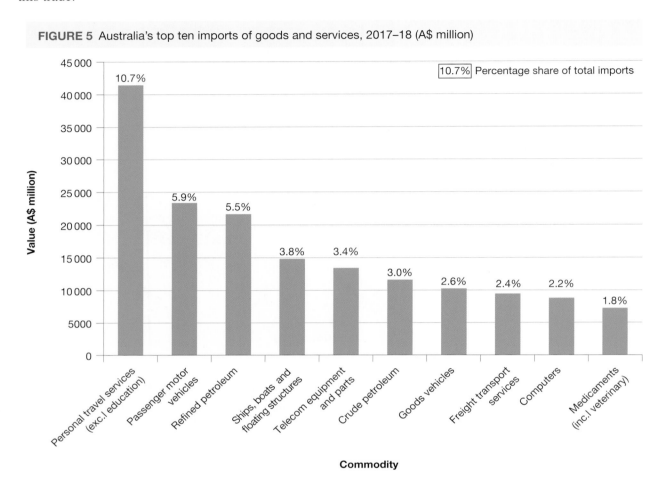

FIGURE 6 Oil and petroleum products make up a significant part of Australia's import trade.

Explore more with myWorldAtlas

Deepen your understanding of this topic with related case studies and questions.
- Investigate additional topics > Australia's links with the world > **Aid, migration and trade**

9.3 EXERCISES

Geographical skills key: GS1 Remembering and understanding **GS2** Describing and explaining **GS3** Comparing and contrasting **GS4** Classifying, organising, constructing **GS5** Examining, analysing, interpreting **GS6** Evaluating, predicting, proposing

9.3 Exercise 1: Check your understanding

1. **GS2** What is the *interconnection* between the World Trade Organization and Australia's trade?
2. **GS1** What are Australia's three most important exports?
3. **GS1** What are Australia's three most important imports?
4. **GS5** Refer to **FIGURE 2** and **TABLE 1**. How many of Australia's top ten two-way trading partners are Asian countries?
5. **GS2** What reasons can you suggest for Australia's significant two-way trade with Asian nations?

9.3 Exercise 2: Apply your understanding

1. **GS2** Despite having a relatively small population, Australia has many goods and services to trade. Explain why this might be so.
2. **GS6** How might a *change* in the growth of Australia's population affect the country's agricultural exports?
3. **GS5** Consider Australia's exports (see **FIGURE 3**).
   (a) What evidence is there in this subtopic to confirm the fact that Australia is regarded as mostly a primary industry exporter?
   (b) Are there any figures for export trade that contradict this statement?
4. **GS6** Look at the goods that Australia imports (see **FIGURE 5**). What factors could lead to a *change* in the types of goods imported by the year 2050?
5. **GS6** Why do you think Australia has become such an important exporter of education services?

Try these questions in learnON for instant, corrective feedback. Go to www.jacplus.com.au.

9.4 Food trade around the world

9.4.1 Trade in food production surpluses

The world's population is unevenly distributed across space, as is the quantity of food produced. Some places, such as Australia, produce an abundance of food, while others struggle to produce enough to maintain food security.

Traditionally, food production consisted of hunting and gathering or cropping and herding. Excess food was consumed locally or sent to nearby markets for **barter** or cash. While some 40 per cent (or by some estimates, more) of the world's population is still directly tied to subsistence agriculture, many of the world's highly developed economies produce large surpluses of food specifically for international trade. For instance, Australia's 2018–19 farm production is estimated to be worth $60 billion, with $47 billion of this in export worth.

The flow of food trade

Much of the flow of food trade is controlled by powerful entities, such as the United States, the European Union and China, and by international food trade agreements. The World Trade Organization (WTO) and the G20 (a group of 20 developed and powerful nations) have a significant say in the flow of food products around the world, particularly with respect to tariffs and fair-trade rules.

Food trade is a complicated business, as can be seen in **FIGURE 1**. It is estimated that for **developing countries**, three-quarters of exports are agricultural produce. While developed countries may need to import some foods, many actually export as much as they import in agricultural produce. For instance, the United States, Canada and Australia use large farms to produce wheat, and they control 75 per cent of the global export trade in cereals.

FIGURE 1 World trade flows — exports of agricultural products by region, US$ billion

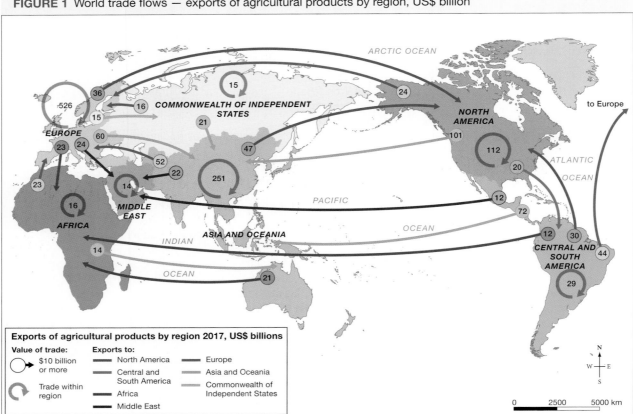

Source: Data from World Trade Organization

Wheat production levels vary from year to year, depending on weather conditions experienced at particular times in the growing season. After a bumper year in 2016–17, when in excess of 34 million tonnes was produced, in 2017–18, Australia produced just over 21 million tonnes of wheat. More than three-quarters of this crop was for sale in overseas markets, worth some $5 billion in export earnings. **FIGURE 2** shows Australia's top ten wheat export partners, with production quantities and earnings averaged over the four-year period up to 2017.

FIGURE 2 Top ten Australian wheat export destinations

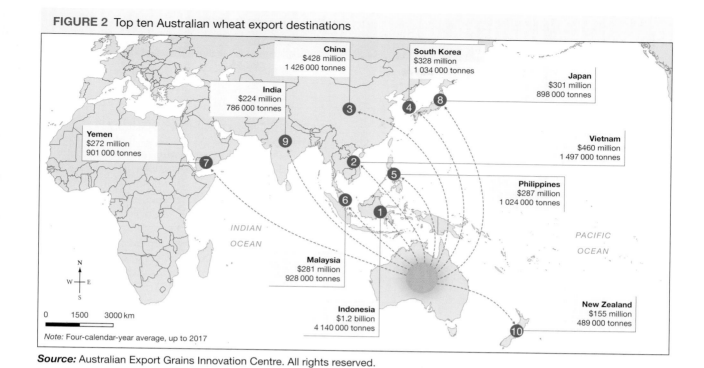

China
$428 million
1 426 000 tonnes

South Korea
$328 million
1 034 000 tonnes

Japan
$301 million
898 000 tonnes

India
$224 million
786 000 tonnes

Yemen
$272 million
901 000 tonnes

Vietnam
$460 million
1 497 000 tonnes

Philippines
$287 million
1 024 000 tonnes

INDIAN
OCEAN

PACIFIC
OCEAN

Malaysia
$281 million
928 000 tonnes

New Zealand
$155 million
489 000 tonnes

Indonesia
$1.2 billion
4 140 000 tonnes

0   1500   3000 km

Note: Four-calendar-year average, up to 2017

FIGURE 3 Australia's wheat exports are worth billions to the economy

Australia's diverse food trade

In addition to wheat, Australia conducts more than $5 billion worth of trade annually through a range of other agriculture, forestry and fisheries exports. **TABLE 1** provides information about Australia's top 20 exports in this category and their respective export values for the 2016–17 financial year.

TABLE 1 Australia's top 20 agriculture, forestry and fisheries exports, 2016–17

Rank	Commodity	$ million	% share
1	Beef	7115	13.8
2	Wheat	6073	11.8
3	Meat (excl. beef)	3832	7.4
4	Vegetables	3270	6.3
5	Wool and other animal hair	3263	6.3
6	Wine	2612	5.1
7	Edible products and preparations	2524	4.9
8	Sugars, molasses and honey	2409	4.7
9	Oil seeds and oleaginous fruits	2278	4.4
10	Barley	2106	4.1
11	Cotton	1787	3.5
12	Fruit and nuts	1769	3.4
13	Live animals (excl. seafood)	1618	3.1
14	Milk, cream, whey and yoghurt	1239	2.4
15	Animal feed	1148	2.2
16	Wood (in chips or particles)	1085	2.1
17	Crustaceans	988	1.9
18	Cheese and curd	848	1.6
19	Hides and skins, raw (excl. furskins)	777	1.5
20	Cereal preparations	728	1.4
Total agriculture, forestry and fisheries exports		**51 643**	

9.4.2 Trade in animals for food

World trade in animals as food is estimated at close to 50 million animals per year — pigs, cattle, goats and sheep. Using modern shipping methods, many animals are transported over long distances, and questions have been asked about potential cruelty in the operation of this trade.

The value of the export trade in animal products from Australia was close to $20 billion in 2016–17. Of this figure, live animal exports comprised $1.6 billion, or 8 per cent. The industry employs up to 10 000 workers at abattoirs, ports and in the transport industry. While there may be concerns about this industry, it should be remembered that some countries request live animal exports so that they can be slaughtered according to **halal** religious customs.

Due to the extensive nature of cattle and sheep farming in Australia, these animals must often travel very large distances to reach ports. They then travel by ship to distant markets. The Australian government has set high standards in the handling of live animals and is monitoring carefully how they are treated at destination ports.

Resources

🔗 **Weblink** Trade

9.4 INQUIRY ACTIVITIES

1. Use the **Trade** weblink in the Resources tab to investigate Australia's top five trade partnerships. Prepare a mind map to show the different goods and services traded with these countries.

Classifying, organising, constructing

2. Investigate the issue of live animal exports from Australia. What are the concerns of those who seek to have live exports stopped? How might a ban on live animal exports from Australia affect farmers? Write two letters to the editor outlining the views of:
   (a) someone who supports live animal exports
   (b) someone who wants to see them banned.

Examining, analysing, interpreting
[Critical and Creative Thinking Capability]

9.4 EXERCISES

Geographical skills key: GS1 Remembering and understanding **GS2** Describing and explaining **GS3** Comparing and contrasting **GS4** Classifying, organising, constructing **GS5** Examining, analysing, interpreting **GS6** Evaluating, predicting, proposing

9.4 Exercise 1: Check your understanding

1. **GS1** What percentage of Australia's total agricultural production is exported?
2. **GS1** What are Australia's five biggest agricultural export products?
3. **GS5** Refer to **FIGURE 1**.
   (a) What is the value of food trade from Oceania to Europe?
   (b) What is the value of food trade from Europe to Oceania?
   (c) Is there a balance in this food trade based on your calculations in parts (a) and (b)?
4. **GS5** Rank the regions of the world in decreasing order by volume of food trade.
5. **GS1** What proportion of Australia's export trade in animal products do live animal exports comprise?

9.4 Exercise 2: Apply your understanding

1. **GS2** Why do countries in *places* such as the Middle East and Asia have a preference for live animal imports?
2. **GS5** Refer to **FIGURE 2**.
   (a) Explain why Australia can export such a large quantity of wheat to the world.
   (b) What reasons can you suggest for why a country such as Russia might not export wheat to Indonesia and Malaysia?
3. **GS5** If the United States is a major trade partner of Australia, why do we not export wheat to the US?
4. **GS6** Suggest ways in which Australia might overcome the problem of drought, which has significant impacts on wheat production tonnage.
5. **GS4** Classify the export commodities in **TABLE 1** according to whether they are animal or non-animal products. Calculate the percentage share of these exports that each category (animal/non-animal) comprises.

Try these questions in learnON for instant, corrective feedback. Go to www.jacplus.com.au.

9.5 SkillBuilder: Constructing multiple line and cumulative line graphs

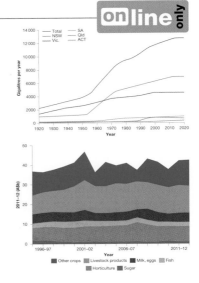

What are multiple line graphs and cumulative line graphs?

Multiple line graphs consist of a number of separate lines drawn on a single graph. Cumulative line graphs are more complex to read, because each set of data is added to the previous line graph.

Select your learnON format to access:

- an overview of the skill and its application in Geography (Tell me)
- a video and a step-by-step process to explain the skill (Show me)
- an activity and interactivity for you to practise the skill (Let me do it)
- questions to consolidate your understanding of the skill.

 Resources

 Video eLesson   Constructing multiple line and cumulative line graphs (eles-1740)

Interactivity   Constructing multiple line and cumulative line graphs (int-3358)

9.6 Impacts of globalisation

9.6.1 Changing trends

Today, you might purchase a jacket online that was designed in Milan, but it is woven from New Zealand wool and stitched together in China. The globalised economy that has resulted from technological developments since the 1990s has brought global marketing, encouraging consumers everywhere to buy goods without considering where they come from. Online shopping has revolutionised the business world by making just about anything imaginable available at the simple tap of a finger or click of a mouse.

The Australian clothing manufacturing industry has produced some very recognisable brand names and distinctive products. Today, the industry faces tough international competition, especially from producers in developing countries who can afford to mass-produce clothing far more cheaply than Australian companies can. As a result, Australian clothing manufacturers tend to focus on high-end, high-quality products rather than attempting to compete with lower-cost producers.

It is not just the clothing industry that has felt the impact of an increasingly globalised economy. Many multinational companies have 'offshored' various production and service divisions to developing countries, such as India, China, Malaysia and the Philippines, due to these countries' lower labour costs. A range of other economically appealing factors, such as a lack of labour unions and incentives offered by those governments including tax breaks and low import duties, have also fueled this trend.

FIGURE 1 This symbol signifies that a product has been manufactured in Australia by an Australian-owned company.

The Global Services Location Index ranks the top destinations for global offshoring. To develop this index three main categories are identified: financial attractiveness, people skills and availability, and business environment. 'Financial attractiveness' constitutes 40 per cent of the index, 'people skills and availability' 30 per cent, and 'business environment' also 30 per cent of the total weighting.

Foreign companies in China

As an example of the growth in global business operations, in 1979, there were 100 foreign-owned enterprises in China. In 1998, there were 280 000, and by the end of 2015, there were more than 835 000 companies with foreign direct investment registered in mainland China. Since 2007, foreign companies have employed more than 25 million people in China. These companies include Coca-Cola, Pepsi, Nike, Citibank, General Motors, Philips, Ikea, Microsoft and Samsung. China's economy is growing rapidly; the country is destined to remain an engine for global growth for some years to come.

FIGURE 2 Top 20 locations for offshore companies, 2017, by region

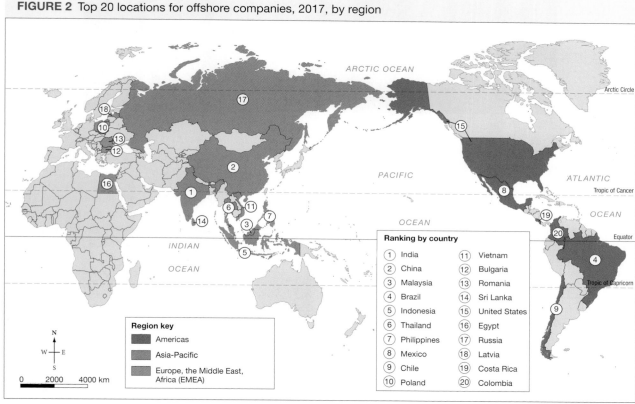

Source: Data from Statista. Map drawn by Spatial Vision.

9.6.2 Sweatshops — a negative side of global trade

If you buy well-known global brands, then you may be wearing clothing or footwear that was made in a sweatshop.

A sweatshop is any working environment in which the workers experience long hours, low wages and poor working conditions. Typically, they are workshops that manufacture goods such as clothing. Sweatshops are common in developing countries, where labour laws are less strict or are not enforced at all. Workers often use dangerous machinery in cramped conditions and can even be exposed to toxic substances. In the worst cases, child labour may be used. Sweatshop workers' wages are generally insufficient to sustain reasonable living conditions; many workers live in poverty. Most are young women aged 17 to 24.

In our globalised world, the question of ethical trade is increasingly important. Socially responsible companies are taking steps to ensure that profits gained from offshoring production and services to less developed countries do not come at the expense of the wellbeing of the people within those countries.

FIGURE 3 A sweatshop in Bangladesh

9.6 INQUIRY ACTIVITY

If clothing carries the Ethical Clothing Australia (ECA) label, it means the garment was manufactured in Australia and the manufacturer has ensured that all people involved in its production received the legally stated wage rates and conditions — known in Australia as award wages and conditions. Research which Australian-made garments you can purchase to support fair working conditions. **Examining, analysing, interpreting**

9.6 EXERCISES

Geographical skills key: GS1 Remembering and understanding **GS2** Describing and explaining **GS3** Comparing and contrasting **GS4** Classifying, organising, constructing **GS5** Examining, analysing, interpreting **GS6** Evaluating, predicting, proposing

9.6 Exercise 1: Check your understanding

1. **GS1** Why have many countries moved their production to offshore *places*?
2. **GS1** What are sweatshops?
3. **GS2** What *change* do you think online shopping will make to the Australian retail industry?
4. **GS3** Compare the advantages and disadvantages of ordering a T-shirt online rather than buying it in a department store.
5. **GS1** What are the three main categories considered in developing the Global Services Location Index?

9.6 Exercise 2: Apply your understanding

1. **GS6** Look at **FIGURE 2**. Suggest reasons why so many offshore manufacturing companies are located in the Asia-Pacific region.
2. **GS5** What impact does moving production offshore have on the Australian economy and people?
3. **GS6** Are sweatshops ethical or *sustainable*? Explain your answer.
4. **GS4** Online ordering of goods is a feature of the internet age. List the advantages and disadvantages of online ordering for workers in the Australian retail industry.
5. **GS5** Australia has made stronger regional trade *interconnections* with its neighbours by lowering its tariffs on imported textiles, clothing and footwear. Suggest two benefits and two disadvantages of this approach for Australian trade and the economy.

Try these questions in learnON for instant, corrective feedback. Go to www.jacplus.com.au.

9.7 Making trade fair

9.7.1 Problems of trade

The benefits of international trade are not evenly shared around the world, and trade often favours developed countries rather than developing countries. It is the role of governments, organisations and agencies to regulate this trade so that the economic benefits are more evenly distributed.

Australians benefit economically, culturally and politically from international trade, but **social justice** problems can arise through this trade. For example, if we import 'blood diamonds' from Africa, clothing manufactured in sweatshops in Bangladesh, or carpets from Nepal produced by child labour, we are supporting unethical industries.

In addition, some countries can make it difficult for other countries to compete fairly, on a 'level playing field'. They do this by:
- *imposing tariffs* — taxes on imports
- *imposing quotas* — limits on the quantity of a good that can be imported
- *providing subsidies* — cash or tax benefits for local farmers or manufacturers.

9.7.2 Fair trade

The fair trade movement aims to improve the lives of small producers in developing nations by paying a fair price to artisans (craftspeople) and farmers who export goods such as handicrafts, coffee, cocoa, sugar, tea, bananas, cotton, wine and fruit. The movement operates through various national and international organisations such as the World Fair Trade Organization and Fairtrade International.

FIGURE 1 Fair trade organisations promote fair labour practices such as preventing and eliminating child labour.

FIGURE 2 Goods produced by workers for the World Fair Trade Organization mission

The fair trade labelling system is operated by Fairtrade International, of which Australia is a participating member. This system works to ensure that income from the sale of products goes back directly to the farmers, artisans and their communities. Fairtrade International works with 1599 producer groups across 75 countries (see **FIGURE 3**). The number of Fairtrade International farmers and workers is estimated at more than 1.6 million, of which 25 per cent are women.

Fairtrade food items include sugar, chocolate, coffee, tea, wine and rice. Other products include soaps, candles, clothing, jewellery, bags, rugs, carpets, ceramics, wooden handicrafts, toys and beauty products.

In 2017–18, Australia and New Zealand had a combined retail sales total of A$333 million in Fairtrade-certified products, with three in five New Zealanders and two in five Australians purchasing Fairtrade offerings. This included 3 million kilograms of coffee, 10.1 million kilograms of chocolate and 354 000 kilograms of Fairtrade tea. On a global scale, Fairtrade's 1.6 million farmers and their families have benefited from Fairtrade premium-funded infrastructure and community development projects with a value of A$262 million.

FIGURE 3 Fairtrade in the world, 2017

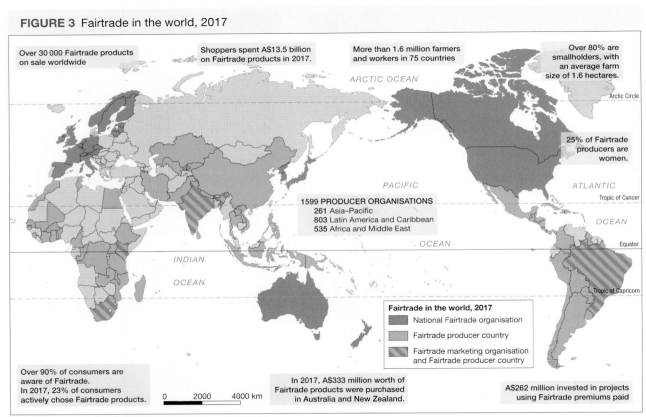

Over 30 000 Fairtrade products on sale worldwide

Shoppers spent A$13.5 billion on Fairtrade products in 2017.

More than 1.6 million farmers and workers in 75 countries

Over 80% are smallholders, with an average farm size of 1.6 hectares.

25% of Fairtrade producers are women.

1599 PRODUCER ORGANISATIONS
261 Asia–Pacific
803 Latin America and Caribbean
535 Africa and Middle East

Fairtrade in the world, 2017
National Fairtrade organisation
Fairtrade producer country
Fairtrade marketing organisation and Fairtrade producer country

Over 90% of consumers are aware of Fairtrade.
In 2017, 23% of consumers actively chose Fairtrade products.

In 2017, A$333 million worth of Fairtrade products were purchased in Australia and New Zealand.

A$262 million invested in projects using Fairtrade premiums paid

0   2000   4000 km

Source: Fairtrade Foundation

9.7.3 Non-government organisations and fair trade

Non-government organisations (NGOs) such as Oxfam and World Vision also support fair trade, and oppose socially unjust trade agreements. They oppose attempts by developed countries to:
- block agricultural imports from developing countries
- subsidise their own farmers while demanding that poorer developing countries keep their agricultural markets open.

9.7 INQUIRY ACTIVITIES

1. Visit your local supermarket and find as many products as you can that carry the Fairtrade symbol. Take pictures of these products and create an annotated world map to show each of the products and where it is produced.   **Classifying, organising, constructing**
2. Conduct internet research to find out what Oxfam does to promote fair trade. What types of goods does Oxfam sell in Australia?   **Examining, analysing, interpreting**

9.7 EXERCISES

Geographical skills key: GS1 Remembering and understanding **GS2** Describing and explaining **GS3** Comparing and contrasting **GS4** Classifying, organising, constructing **GS5** Examining, analysing, interpreting **GS6** Evaluating, predicting, proposing

9.7 Exercise 1: Check your understanding

1. **GS1** What are the main principles of fair trade?
2. **GS1** List three products that are sold under the Fairtrade International banner.
3. **GS1** How many farmers throughout the world are associated with Fairtrade International?
4. **GS2** Explain the role of NGOs such as Oxfam in relation to trade.
5. **GS2** Why can trade be unfavourable to poorer countries?

9.7 Exercise 2: Apply your understanding

1. **GS5** Describe the distribution of Fairtrade producer countries.
2. **GS5** In which parts of the world are National Fairtrade Organisation countries found?
3. **GS6** How could awareness of the work of Fairtrade be increased?
4. **GS2** Explain how consumers in developed countries may unwittingly support unethical enterprises.
5. **GS6** In theory, every country, rich or poor, should have the opportunity to benefit from international trade. However, the reality is very different. Write a page discussing this statement.

Try these questions in learnON for instant, corrective feedback. Go to www.jacplus.com.au.

9.8 Global connections through Australian Aid

9.8.1 Why give overseas aid?

Overseas aid is the transfer of money, food and services from developed countries such as Australia to less-developed countries in order to help people overcome poverty, resolve humanitarian issues and generally help with their development. Over one billion people in the world live in poverty and do not have easy access to education and health care. When disasters strike, they lack the resources to get back on their feet. Poverty needs to be addressed by the international community because it can:
- breed instability and **extremism**
- cause people to flee violence and hardship, thus swelling the number of refugees.

Australia takes the stance that helping people who are less fortunate is a vital way of supporting **humanitarian principles** and social justice. Apart from showing we care, it is in the interests of our **national security** as it may also help promote stability and prosperity in the region. In addition, it improves our status throughout the world and creates political and economic interconnections with our Asia-Pacific neighbours. Australia's Official Development Assistance (ODA) program is known as Australian Aid.

9.8.2 The Australian Aid program

The Department of Foreign Affairs and Trade (DFAT) manages the Australian government's multi-billion-dollar overseas aid program. To ensure that funds reach those in need, Australian Aid works with Australian businesses, non-government organisations such as CARE Australia, and international agencies such as the United Nations (UN) and the World Bank. In 2018–19, Australia's ODA budget was $4.2 billion, with the majority of this being earmarked for the Indo-Pacific region, of which Australia is a part (see **FIGURE 1**).

There are various investment priorities within Australia's ODA budget (see **FIGURE 2**). Within these priorities, many programs that target specific areas of need or interest are covered. These include:

- aid to governments for post-conflict reconstruction, as in Afghanistan
- distribution of food through the United Nations World Food Programme
- contributions to United Nations projects on refugees and climate change
- disaster and conflict relief in the form of food, medicine and shelter
- programs by non-government organisations to reduce child labour in developing countries
- funding for education programs
- funding for programs to promote gender equality and improve women's economic and social participation
- support for Australian volunteers working overseas.

FIGURE 1 Australia's aid 2018–19, by region

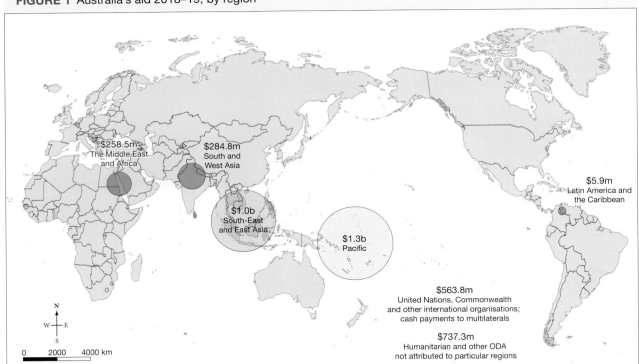

Source: Data from © Commonwealth of Australia, DFAT, Australian Aid Budget Summary 2018–19. Map drawn by Spatial Vision.

DISCUSS

Australia should help its less developed neighbours, not just because it benefits Australia but because it is the right thing to do. **[Ethical Capability]**

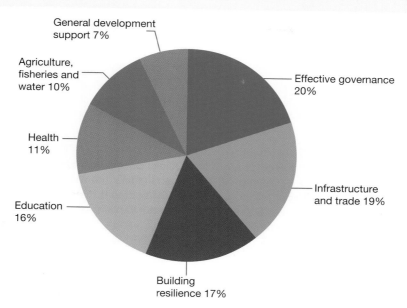

The Australian Aid program supports the United Nations Sustainable Development Goals (SDGs). In addition, countries with a low Human Development Index (HDI) score are the target for development assistance. The HDI ranks countries according to life expectancy, education and per capita income. The highest possible score for a country is 1.0; countries with low HDI ranking score below 0.55 (see **FIGURE 3**). Australian ODA aims to improve the lives of people in such countries through programs and initiatives that seek to build social and economic resilience.

FIGURE 3 The Human Development Index, 2017

Source: Data from UNDP Human Development Reports. Map drawn by Spatial Vision.

9.8 EXERCISES

Geographical skills key: GS1 Remembering and understanding **GS2** Describing and explaining **GS3** Comparing and contrasting **GS4** Classifying, organising, constructing **GS5** Examining, analysing, interpreting **GS6** Evaluating, predicting, proposing

9.8 Exercise 1: Check your understanding

1. **GS1** Which government department manages Australia's Official Development Assistance (ODA) program?
2. **GS2** Which regions of the world receive most of Australia's aid funding and why do you think this is so?
3. **GS1** List the different focus areas across which Australia's ODA budget is distributed.
4. **GS2** Is there a case that could be argued for cutting aid budgets? Explain your reasons.
5. **GS2** Describe the global distribution of low-HDI countries.

9.8 Exercise 2: Apply your understanding

1. **GS6** If Australian Aid was to stop, what impact do you think this would have on Australia's reputation in the international community?
2. **GS6** Which elements of the Australian Aid program do you think will have the greatest impact on the lives of people in the Pacific region? Give reasons for your selection.
3. **GS3** Compare the distribution of low-HDI countries with that of high-HDI countries. Suggest an explanation for what you observe.
4. **GS6** If Australia's economic prosperity were to decline in the next 50 years, which elements of the Australian Aid program do you believe would not be **sustainable**?
5. **GS5** What reasons can you put forward to explain why Australian Aid programs are worthwhile in terms of Australia's **interconnections** with its neighbours?

Try these questions in learnON for instant, corrective feedback. Go to www.jacplus.com.au.

9.9 The troubling illegal wildlife trade

9.9.1 Threatened wildlife

Not all trade is legal. The international trade in wildlife has been one of the factors responsible for the decline in many species of animals and plants (see **FIGURES 1** and **2**). Millions of live birds, reptiles, mammals, insects and plants are illegally shipped around the world each year (see **FIGURE 3**). Some will supply the pet trade, others may be used in traditional medicines or as food delicacies. Wild animal and plant products, such as skins, meat, ornaments, animal parts and timber, are traded in enormous quantities — estimated to be worth US$10 billion per year.

9.9.2 What is traded and how?

Trade in threatened wildlife takes place through **smuggling**. Birds are drugged and stuffed into plastic tubes; snakes are coiled into stockings and posted; lizards are stitched into suitcases. Many of the animals die.

Prices on the **black market** can be very high. Bird traffickers can earn more than $150 000 for taking 30 eggs out of a country in specially designed vests that keep the eggs warm. Overseas collectors will pay up to $50 000 for a breeding pair of endangered red-tailed black cockatoos.

FIGURE 1 Number of threatened mammal species

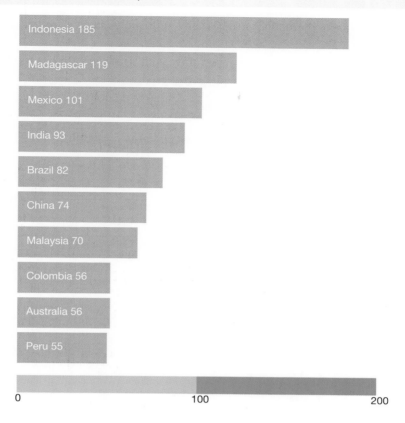

Indonesia 185
Madagascar 119
Mexico 101
India 93
Brazil 82
China 74
Malaysia 70
Colombia 56
Australia 56
Peru 55

0   100   200

FIGURE 2 Number of threatened bird species

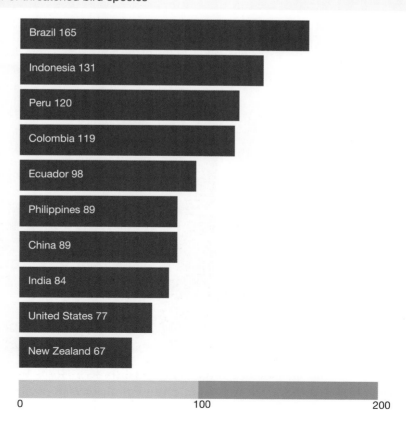

Brazil 165
Indonesia 131
Peru 120
Colombia 119
Ecuador 98
Philippines 89
China 89
India 84
United States 77
New Zealand 67

0   100   200

FIGURE 3 Who trades what — exporters and importers in the illegal wildlife trade

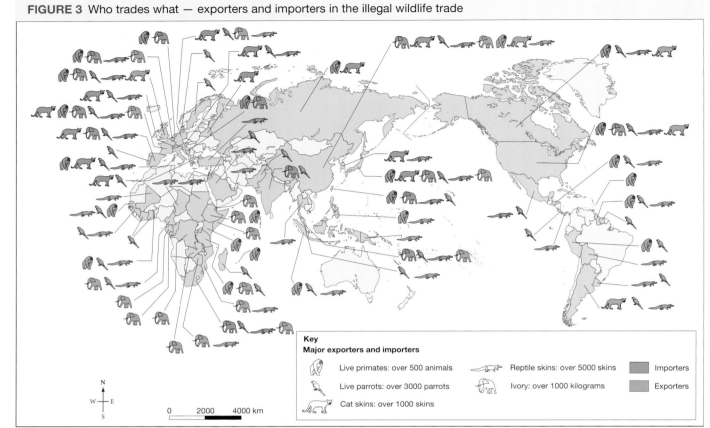

Key
Major exporters and importers

Live primates: over 500 animals	Reptile skins: over 5000 skins
Live parrots: over 3000 parrots	Ivory: over 1000 kilograms
Cat skins: over 1000 skins	

Importers

Exporters

Source: MAPgraphics Pty Ltd, Brisbane

Traditional Chinese medicine

Traditional Chinese medicine (TCM), the most widely practised traditional medicine system in the world, uses more than 1000 plant and animal species. While TCM has been practised for perhaps 5000 years, some of the wild plants and animals used are now threatened or in danger of extinction. Among them are certain orchids, musk deer, rhinoceroses, tigers and some bear species.

All five species of rhinoceros are threatened with extinction due to a long history of being killed by poachers who then remove and sell their horns. The black rhino horn is sometimes referred to as 'black gold' because it is so expensive. Rhinoceros horns are mostly used in traditional Asian medicines to treat a variety of ailments, including fever and blood disorders. Horns are cut into oblong pieces and smuggled into other countries in jars of honey, cartons of matches or raw meat.

FIGURE 4 Some of the skins seized from poachers in India

9.9.3 Tackling the issue

In 1973, an international **treaty** known as the Convention on International Trade in Endangered Species (CITES) was established to prevent international trade that threatened species with extinction. Any trade in products from threatened wildlife requires a special permit. Unless there are exceptional circumstances, no such permits are issued for species threatened with extinction. These include species of tigers, elephants, rhinoceroses, apes, parrots and all sea turtles.

Although international trade in rhino horn has been banned since 1977, demand remains high, which encourages rhino poaching in Africa and Asia. Criminal syndicates link the poachers, in places like South Africa, to transit points, smuggling channels and final destinations in Asia. To combat this, conservationists sedate rhinos in the wild and remove their horns, thus removing the commodity that poachers seek in killing them.

FIGURE 5 Conservationists dehorning a sedated rhinoceros

Creating wildlife reserves and employing rangers to patrol and protect the wildlife within them has had some success, but poaching continues, and in some cases the lives of the rangers themselves are at risk from criminal poachers focused solely on financial gain. In 2017, 100 rangers died in the course of their work.

In October 2018, the Illegal Wildlife Trade Conference was held in London. At this event, wildlife conservationists, policymakers and others sought to coordinate efforts to put an end to criminal activity in the trade of wildlife — an ongoing problem that threatens plants, animals, humans, and the planet's biodiversity alike. While awareness has grown, and concerted measures are being undertaken, there is still much to be done to bring this trade to a halt.

9.9 INQUIRY ACTIVITY

Look carefully at **FIGURE 3**.
a. Which categories of wildlife are traded for their skins?
b. Using an atlas, list the **places** that are the major exporters of live primates, cat skins, ivory, live parrots and reptile skins.
c. Based on your list in part b, list the continents that are the main sources of wildlife species and wildlife goods.
Classifying, organising, constructing

9.9 EXERCISES

Geographical skills key: GS1 Remembering and understanding **GS2** Describing and explaining **GS3** Comparing and contrasting **GS4** Classifying, organising, constructing **GS5** Examining, analysing, interpreting **GS6** Evaluating, predicting, proposing

9.9 Exercise 1: Check your understanding

1. **GS1** Why are wild species traded?
2. **GS1** Which three countries have the highest number of threatened mammal species?
3. **GS1** Which three countries have the highest number of threatened bird species?
4. **GS2** Refer to **FIGURE 5**.
   (a) How does the action shown support the *sustainability* of rhinoceros populations?
   (b) Why are conservationists taking this action?
5. **GS2** What was the purpose of the 2018 Illegal Wildlife Trade Convention?

9.9 Exercise 2: Apply your understanding

1. **GS6** Other than the loss of endangered species, what other negative consequences may occur through the illegal export and import of wildlife?
2. **GS6** What might be the impact on the Australian *environment* if animals such as koalas were to become a target for smugglers?
3. **GS6** Suggest what could be done to persuade people not to buy poached products such as animal skins.
4. **GS6** Propose ways of educating people to move away from the use of traditional medicines that endanger wildlife species.
5. **GS6** Suggest some strategies to reduce the number of wildlife rangers who are killed in the course of their work.

Try these questions in learnON for instant, corrective feedback. Go to www.jacplus.com.au.

9.10 SkillBuilder: Constructing and describing a flow map

What is a flow map?

A flow map is a map that shows the movement of people or objects from one place to another. Arrows are drawn from the point of origin to the destination. Sometimes these lines are scaled to indicate how much of the feature is moving. Thicker lines show a larger amount; thinner lines show a smaller amount.

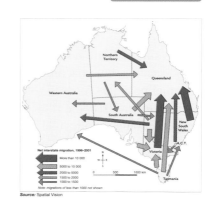

Select your learnON format to access:

- an overview of the skill and its application in Geography (Tell me)
- a video and a step-by-step process to explain the skill (Show me)
- an activity and interactivity for you to practise the skill (Let me do it)
- questions to consolidate your understanding of the skill.

On Resources

Video eLesson Constructing and describing a flow map (eles-1741)

Interactivity Constructing and describing a flow map (int-3359)

9.11 Thinking Big research project: World Trade Fair infographic

SCENARIO

The Department of Foreign Affairs and Trade is running a competition looking for exciting, engaging infographic posters depicting Australia's export and import trade relationships, to be displayed at the upcoming World Trade Fair. Winners of 'Best poster' will get to present their infographic at the fair.

Select your learnON format to access:

- the full project scenario
- details of the project task
- resources to guide your project work
- an assessment rubric.

 Resources

 projectsPLUS Thinking Big research project: World Trade Fair infographic (pro-0195)

9.12 Review

9.12.1 Key knowledge summary

Use this dot point summary to review the content covered in this topic.

9.12.2 Reflection

Reflect on your learning using the activities and resources provided.

 Resources

 eWorkbook   Reflection (doc-31728)

Crossword (doc-31729)

 **Interactivity**   Trade — a driving force for interconnection crossword (int-7651)

KEY TERMS

barter to trade goods in return for other goods or services rather than money

black market any illegal trade in officially controlled or scarce goods

developing countries nations with a low living standard, undeveloped industrial base and low human development index relative to other countries

extremism extreme political or religious views or extreme actions taken on the basis of those views

halal describes food that is prepared under Islamic dietary guidelines

humanitarian principles the principles governing our response to those in need, with the main aim being to save lives and alleviate suffering

national security the protection of a nation's citizens, natural resources, economy, money, environment, military, government and energy

non-government organisation (NGO) a group or business that is organised to serve a particular social purpose at local, national or international level, and operates independently of government

offshore to relocate part of a company's processes or services overseas in order to decrease costs

smuggling importing or exporting goods secretly or illegally

social justice a principle applied so that a society is based on equality, the appreciation of the value of human rights and the recognition of the dignity of every human being

trade barrier government-imposed restriction (in the form of tariffs, quotas and subsidies) on the free international exchange of goods or services

trading partner a participant, organisation or government body in a continuing trade relationship

treaty a formal agreement between two or more independent states or nations, and usually involving a signed document

value adding processing a material or product and thereby increasing its market value

10 Global ICT — connections, disparity and impacts

10.1 Overview

Technology makes our lives easier and helps us connect, but at what cost to people and the planet?

10.1.1 Introduction

The rapid development of information and communication technologies has led to immense change in the way we live and work, and has created a degree of global connectedness unseen and perhaps unimaginable in the past. But not everyone is equally connected. Variations in access exist that lead to social and economic disparity, and there are social, economic and environmental impacts of our increasing production and consumption of technology-related goods and the **e-waste** that results.

 Resources

☑ **eWorkbook**　　Customisable worksheets for this topic

🎞 **Video eLesson**　Plugging in (eles-1725)

LEARNING SEQUENCE

10.1 Overview
10.2 Information and communications technology
10.3 The internet connects us
10.4 Connected Australians
10.5 Improving lives via digital connection — Kenya
10.6 Forging new ICT directions — India
10.7 The impact of ICT production
10.8 The future for e-waste
10.9 **SkillBuilder:** Constructing a table of data for a GIS　　`online only`
10.10 **SkillBuilder:** Using advanced survey techniques — interviews　　`online only`
10.11 **Thinking Big research project:** Trash or treasure?　　`online only`
10.12 **Review**　　`online only`

To access a pre-test and starter questions and receive immediate, **corrective feedback** and **sample responses** to every question, select your learnON format at www.jacplus.com.au.

10.2 Information and communications technology

10.2.1 Changing communications technology

The information and communications technology (ICT) sector is a rapidly evolving aspect of our lives. Change is ongoing, with new technologies constantly emerging. At the same time, some technologies have been superseded. **FIGURE 1** shows the surge in use of mobile phones and in particular the active (used within last 30 days) use of mobile broadband, compared to the decline of the fixed telephone line.

FIGURE 1 The change in our use of technology

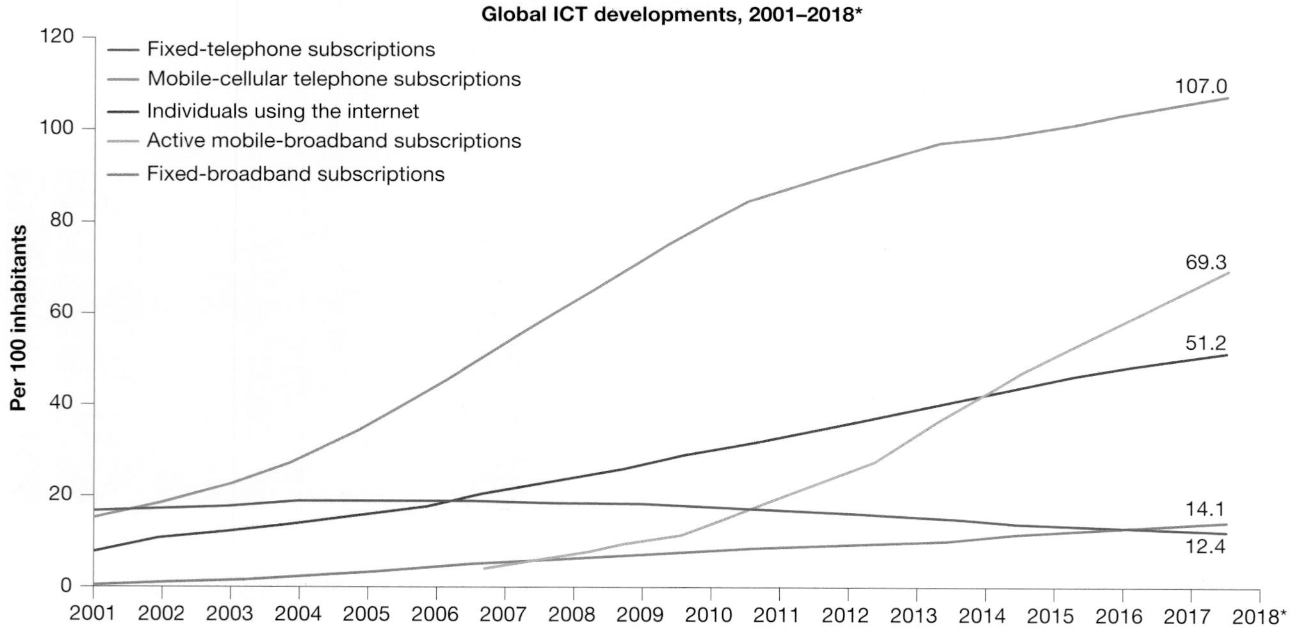

Global ICT developments, 2001–2018*

Legend:
- Fixed-telephone subscriptions
- Mobile-cellular telephone subscriptions
- Individuals using the internet
- Active mobile-broadband subscriptions
- Fixed-broadband subscriptions

Y-axis: Per 100 inhabitants

Data labels: 107.0, 69.3, 51.2, 14.1, 12.4

X-axis: 2001 2002 2003 2004 2005 2006 2007 2008 2009 2010 2011 2012 2013 2014 2015 2016 2017 2018*

Note: *Estimate

Source: ITU World Telecommunication/ICT Indicators database

The **World Wide Web** was developed as a way of accessing and spreading information. It was once simply a means of collaboration and exchanging ideas online. Today it is an enabler that makes our lives connected to almost everything through the internet.

The first mobile phones in the 1980s were used solely for conversation. Today mobile phones have evolved with a global demand for smartphones — technology that can map travel routes, take photos and videos, act as a diary or notebook, do shopping and banking, participate in gaming, record music, print documents wirelessly, allow face-to-face talking, share documents via the cloud and much, much more. Applications (apps) are being developed at a high rate for the interpretation and use of everything from human health matters, to bird calls, to alerts for disaster management, and so on. Virtual reality is taking us places we have never been.

Although there are more than 750 million adults in the world who lack basic literacy skills, youth culture worldwide has adopted ICT as a mainstream part of life. It has become a fundamental element in the way many of us connect to services and information, and to people in other places. Today, globally, there are more people using the internet on their mobile phones than those using the internet from a stand-alone computer (see **FIGURE 2**).

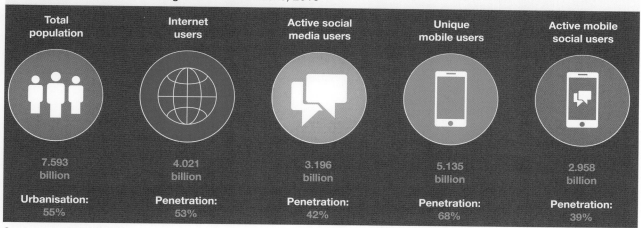

FIGURE 2 Global users of digital communications, 2018

Total population	Internet users	Active social media users	Unique mobile users	Active mobile social users
7.593 billion	4.021 billion	3.196 billion	5.135 billion	2.958 billion
Urbanisation: 55%	Penetration: 53%	Penetration: 42%	Penetration: 68%	Penetration: 39%

Source: wearesocial.com

10.2 EXERCISES

Geographical skills key: GS1 Remembering and understanding **GS2** Describing and explaining **GS3** Comparing and contrasting **GS4** Classifying, organising, constructing **GS5** Examining, analysing, interpreting **GS6** Evaluating, predicting, proposing

10.2 Exercise 1: Check your understanding

1. **GS1** What was the initial purpose of the World Wide Web (www)?
2. **GS2** What does it mean to say the www is an 'enabler'?
3. **GS1** List how the uses of mobile phones are different to when they were introduced in the 1980s.
4. **GS1** How many adults in the world lack basic literacy skills?
5. **GS2 FIGURE 2** shows the global users of ICT. Describe the role of mobile phones in our lives.

10.2 Exercise 2: Apply your understanding

1. **GS2** Using **FIGURE 1**, describe the *change* over time from 2001 to 2018 of the technologies shown.
2. **GS6** Suggest the innovations in ICT that have *changed* people's use of the internet.
3. **GS6** Smartphones have taken communications to a 'new level'. What might smartphones or the next generation of phones be able to do in the future?
4. **GS6** Suggest why it is youth culture that has adopted technology so readily into their lives.
5. **GS6** Will computers become extinct for communications in the future? Explain your view.

Try these questions in learnON for instant, corrective feedback. Go to www.jacplus.com.au.

10.3 The internet connects us

10.3.1 Global internet connections

Internet **connectivity**, whether via a computer or a mobile phone, is available across the world, but its distribution is not even across regions or within countries. From **FIGURE 1** it is clear that the regions with a very high level of **human development**, for example Europe and North America, also have a high level of internet users. The countries of Middle and Eastern Africa with a lower level of human development have fewer people using the internet. **TABLE 1** shows the **digital divide** between countries and regions, with the top ten countries and the bottom ten countries measured per head of population using the internet.

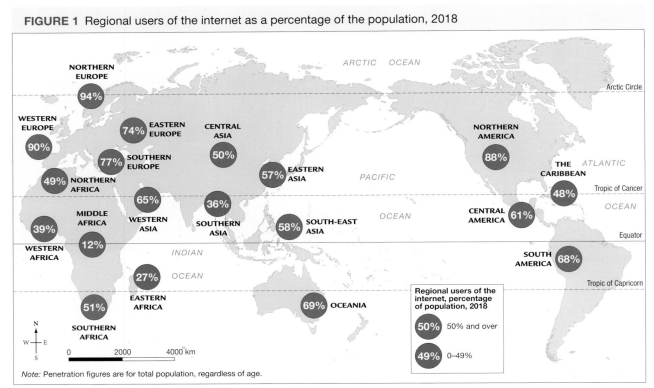

FIGURE 1 Regional users of the internet as a percentage of the population, 2018

Source: Internetworldstats; ITU; Eurostat; Internetlivestats; CIA World Factbook; Mideastmedia.org; Facebook; government officials; regulatory authorities; reputable media

TABLE 1 Countries with the highest and lowest population proportion using the internet, 2018

Rank	Country	Proportion of population	Number of users
Top ten			
1	Qatar	99%	2 640 360
2	United Arab Emirates	99%	9 376 171
3	Kuwait	98%	4 100 000
4	Bermuda	98%	60 125
5	Bahrain	98%	1 499 193
6	Iceland	98%	329 675
7	Norway	98%	5 222 786
8	Andorra	98%	75 366
9	Luxembourg	98%	572 216
10	Denmark	97%	5 571 635
Bottom ten			
213	North Korea	0.06%	16 000
212	Eritrea	1%	71 000

(continued)

TABLE 1 Countries with the highest and lowest population proportion using the internet, 2018 *(continued)*

Rank	Country	Proportion of population	Number of users
211	Niger	4%	946 440
210	Western Sahara	5%	28 000
209	Chad	5%	756 329
208	Central African Republic	5%	246 432
207	Burundi	6%	617 116
206	Democratic Republic of the Congo	6%	5 133 940
205	Guinea-Bissau	6%	120 000
204	Madagascar	7%	1 900 000

10.3.2 Mobile phones connect with the internet

Today people of all ages carry a mobile phone. For young people, it is just a regular way to connect with friends, family and the world. People in their middle years were introduced to the technology as young adults and have embraced the interconnections provided; they readily take on each new development offered by the service providers. For older people, the adaptation to the technology has needed to be rapid and many see the technology as complex — understanding the technology and mastering the skills are a challenge, especially to many people over the age of 80.

Just like internet access, the distribution of mobile phones across the world is not even (see **FIGURE 2**).

FIGURE 2 The distribution of mobile subscriptions per 100 people, 2017

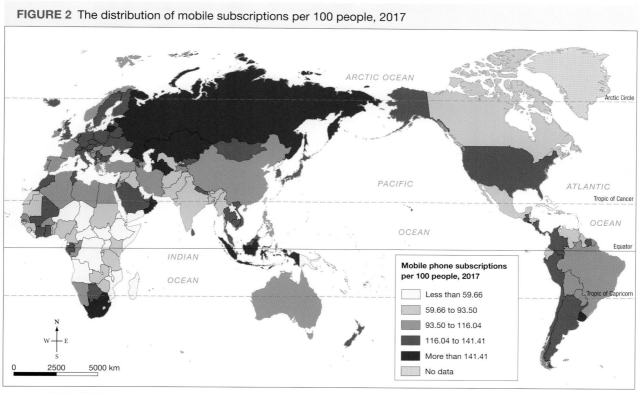

Source: JUMIA (2018)

In countries with a very high level of human development there has been a shift to smartphones. In the countries with a lower level of human development the adoption of the latest technology is not as evident. **FIGURE 3** shows the adoption of mobile phones and smartphones in a few selected countries.

FIGURE 3 The adoption of mobile phones by adults in selected countries, 2018

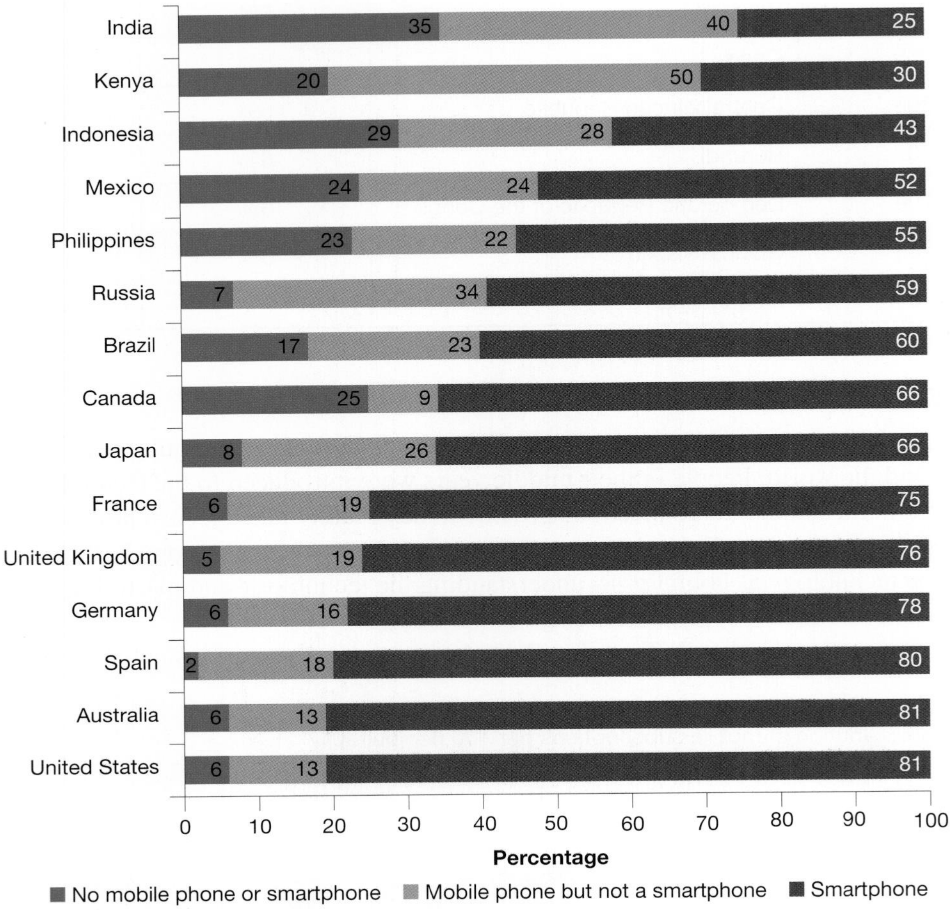

■ No mobile phone or smartphone   ■ Mobile phone but not a smartphone   ■ Smartphone

10.3 INQUIRY ACTIVITIES

1. **GS4** Refer to **FIGURE 1**.
   (a) On a blank map of the world, shade the different regions according to the data shown. Develop a key for 'very high', 'high', 'medium', 'low' and 'very low' users of the internet.
   (b) Compare your map with those of others in your class. Is your map the same as someone else's? Why or why not?
   (c) Describe the pattern of users shown on your map. **Classifying, organising, constructing**
2. **GS5** Using a world map, find the countries listed in **TABLE 1**.
   (a) In which parts of the world is the highest proportion of internet use found? Suggest a reason for this occurrence.
   (b) In which parts of the world is the lowest proportion of internet use found? Suggest reasons for this occurrence. **Examining, analysing, interpreting**

10.3 EXERCISES

Geographical skills key: GS1 Remembering and understanding **GS2** Describing and explaining **GS3** Comparing and contrasting **GS4** Classifying, organising, constructing **GS5** Examining, analysing, interpreting **GS6** Evaluating, predicting, proposing

10.3 Exercise 1: Check your understanding

1. **GS1** Is everyone across the world connected to the internet?
2. **GS1** In which regions of the world is the lowest number of people using the internet?
3. **GS2** Using **TABLE 1**, name three countries in which the proportion of the population using the internet is greater than 97 per cent.
4. **GS1** Which age groups have been able to handle the *changes* in mobile phone connectivity with the internet better than other age groups?
5. **GS2** In which parts of the world has the adoption of smartphones been greatest?

10.3 Exercise 2: Apply your understanding

1. **GS4** Using statistics from **FIGURE 3** to support your answer, describe the level of mobile phone use in:
   (a) India
   (b) Kenya
   (c) Australia.
2. **GS5** Mobile phone use differs between the developed world and the less developed world. Use data from **FIGURE 3** to support this statement.
3. **GS6** Choose one European country, one African country and one Asian country included in **FIGURE 3** and hypothesise the *changes* to mobile phone adoption that might occur in those countries by 2030.
4. **GS6** Suggest three reasons for the uneven distribution of mobile phones across the world.
5. **GS6** For the next generation of mobile phones after smartphones, provide a reasoned answer as to which countries might be the first adopters of such a technology.

Try these questions in learnON for instant, corrective feedback. Go to www.jacplus.com.au.

10.4 Connected Australians

10.4.1 Australia's digital divide

Australia is a highly developed country in which we consider access (immediate connection, advanced equipment, and high data allowances) to the internet a necessity. Australians also expect the technology to be affordable as a proportion of their income. Our ability to adapt to the rapidly changing environment and our high skill levels are such that Australians make good use of their connectivity. However, not everyone across the country has equal access to the internet — there is a 'digital divide', whereby some areas experience greater levels of digital inclusion than others.

The Australian Digital Inclusion Index (ADII) takes into consideration the three key components of ICT quality — access, affordability and digital ability. **FIGURE 1** shows the Australian average at a medium level (rating 56.5 from a possible 100). Most of the states are around that average, although people across South Australia and Tasmania appear to be less well connected. The divide is further evident between the capital cities and the rural areas. According to the 2017 ADII report, Australia's least digitally included regions are: Burnie and western Tasmania (44.1), north-west Queensland (45.9), north Victoria (46.5), east Victoria (47.0), Launceston and north-east Tasmania (47.7), and north-west Victoria (48.2).

10.4.2 Some Australians are less well connected

In addition to disparities in connectedness based on geographical location, there are also particular groups within Australian society that are more digitally disadvantaged. **FIGURE 2** shows that people with lower incomes, those with no income, those older than 50 years and especially those over 65, the disabled and the Indigenous (remote communities were not included in the ADII) have a digital inclusion index lower than the Australian average.

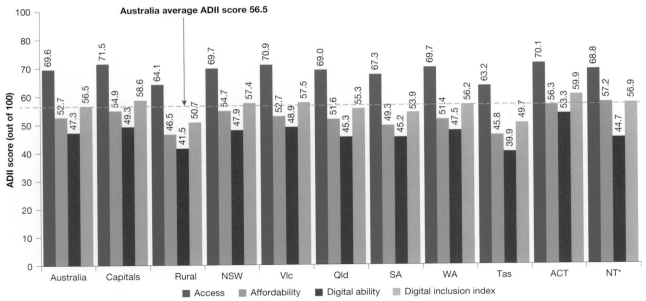

FIGURE 1 Australian Digital Inclusion Index by state, 2017

*Sample size <100; exercise caution in interpretation.

Source: Roy Morgan Research, April 2016–March 2017

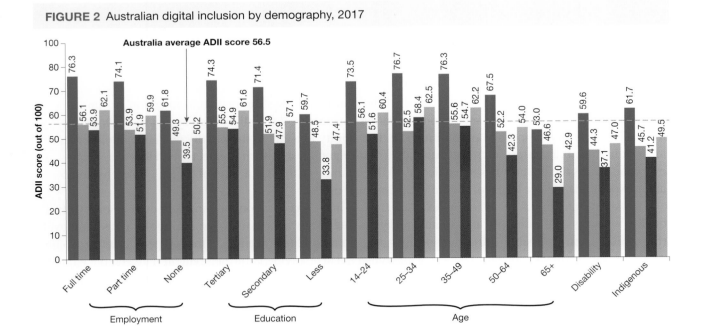

FIGURE 2 Australian digital inclusion by demography, 2017

Source: Roy Morgan Research, April 2016–March 2017

10.5 Improving lives via digital connection — Kenya

10.5.1 Increased consumption of ICT

Globally, there is a difference between the developed and the developing world in terms of levels of digital connection, but it is important to note that access to the internet and mobile phone networks has improved, especially across Africa. In Kenya, for example, many people live in rural and remote places in the countryside. In the past this left families disconnected from one another, as the primary earner in the family often had to work in a distant town to provide the family income. No longer do these rural and remote people have to make a long journey on poor quality roads into town to transfer money. Today these people use their mobile phones.

The number of mobile phone subscribers in Kenya has risen steadily (see **FIGURE 1**) and the mobile phone coverage has spread with the introduction of each new network speed (see **FIGURE 2**). Internet connection is via mobile phones and the young are the dominant users — 52 per cent of the 16–24 age group and 39 per cent of the 25–34 age group, as opposed to just six per cent of the 35–44 group and four per cent of the 45+ group.

FIGURE 1 Mobile phone subscriptions and internet penetration, Kenya

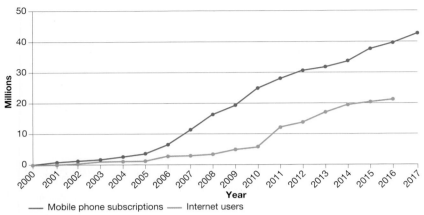

FIGURE 2 Mobile phone network coverage, Kenya, 2017

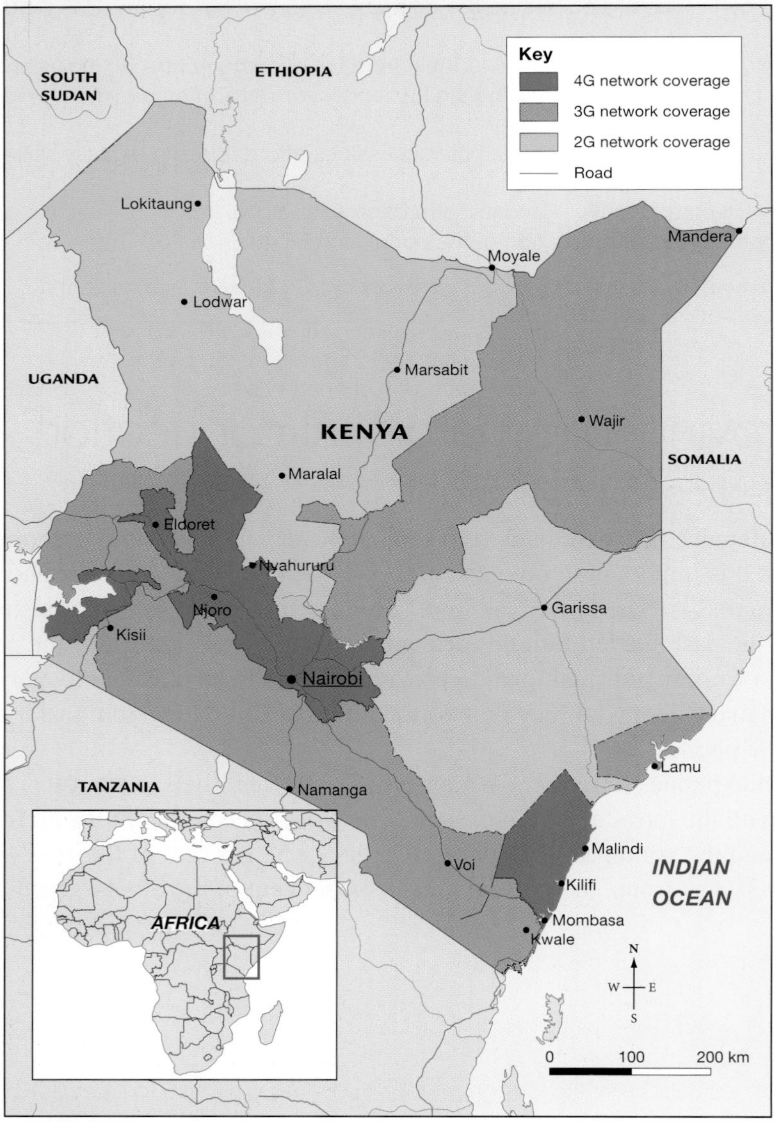

Key
- 4G network coverage
- 3G network coverage
- 2G network coverage
- Road

SOUTH SUDAN

ETHIOPIA

Lokitaung

Moyale

Mandera

Lodwar

Marsabit

UGANDA

KENYA

Wajir

SOMALIA

Maralal

Eldoret

Nyahururu

Garissa

Njoro

Kisii

Nairobi

Lamu

TANZANIA

Namanga

Voi

Malindi

Kilifi

INDIAN OCEAN

Mombasa

Kwale

AFRICA

N
W — E
S

0   100   200 km

Source: JUMIA (2018)

10.5.2 Mobile money and moving forward

In 2007, the UK-based organisation Financial Deepening Challenge Fund (FDCF) worked in Kenya to set up M-Pesa (meaning 'mobile money'), and various agencies were set up to assist users (see **FIGURE 3**). Customers could then transfer, withdraw and deposit money through mobile phones; nearly 50 per cent of Kenya's **gross domestic product** (GDP) is processed over M-Pesa.

FIGURE 3 A customer at an M-Pesa agent

Access to mobile phones for small-business owners has meant they are now able to advertise to a larger audience and are no longer dependent on word-of-mouth advertising. Clients can now contact business operators with ease. For those working away from home, it is a safe and easy way to send money back to families in the countryside. M-Pesa has eliminated the need to carry large sums of cash to markets, thus improving personal safety.

M-Pesa demonstrates how dreaming big but thinking locally can have a significant effect on the economic and social structure of a place, just through the use of a mobile phone.

FIGURE 4 The impact of M-Pesa

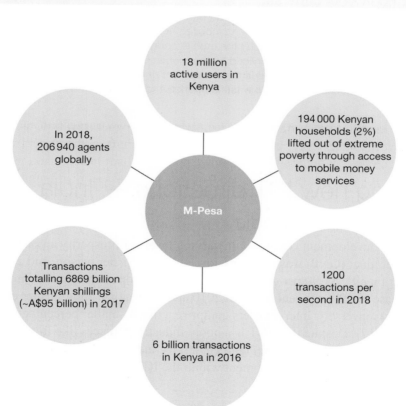

Kenya has 'jumped forward' with its use of ICT — it is now dubbed the Silicon Valley of Africa. About 84 per cent of the population is connected to the internet. Additional service providers have opened up innovative platforms to give Kenyans connection to the world. For example, Kenyan channel KTN News joined YouTube in 2016 with 145 000 uploaded videos and 278 300 subscribers. With increased providers covering 90 per cent of the country, mobile phone packages are more affordable and smartphones are now cheaper, to the benefit of the consumers. By 2018 several local digital start-ups and international ICT companies were calling Kenya home. Kenya is bridging the digital divide.

a. In small groups, suggest a list of possible criteria that you could use to judge how effective technology has been in improving people's lives in Kenya.
b. Share and discuss your group's criteria with the class and select the three criteria that the class considers the most effective for judging.
c. Did you have to make many changes to create the class list? Why or why not? How difficult was it to reach a consensus? **[Critical and Creative Thinking Capability]**

10.5 EXERCISES

Geographical skills key: GS1 Remembering and understanding **GS2** Describing and explaining **GS3** Comparing and contrasting **GS4** Classifying, organising, constructing **GS5** Examining, analysing, interpreting **GS6** Evaluating, predicting, proposing

10.5 Exercise 1: Check your understanding

1. **GS2** Make a list of five *interconnections* than can occur between Kenyans thanks to M-Pesa.
2. **GS2** Use **FIGURE 4** to describe the success rate of M-Pesa. Use specific data in your answer.
3. **GS2** Explain how Kenya is bridging the digital divide.
4. **GS1** Which age group in Kenya is the dominant user of mobile phones?
5. **GS1** What percentage of Kenya is covered by ICT providers?

10.5 Exercise 2: Apply your understanding

1. **GS5** Mobile phone usage has been an important part of the improvement in communications in Kenya. Using **FIGURE 1**, describe the *change* over time in mobile phone subscriptions and users of the internet.
2. **GS5** Which areas of Kenya are well serviced by the mobile phone network?
3. **GS5** Which *places* in Kenya are not well serviced by the mobile phone network?
4. **GS5** Which parts of Kenya have the most recent development of 4G services?
5. **GS2** Is there a correlation between the use of mobile phones and the internet as shown in **FIGURE 1**? Explain your answer.

Try these questions in learnON for instant, corrective feedback. Go to www.jacplus.com.au.

10.6 Forging new ICT directions — India

10.6.1 The digital divide in India

India is a medium-level-development country with varying levels of prosperity among its people. In 2017, Indian gross national product (GNP) was relatively low, at US$6353 per person, and 21.2 per cent of the population earned less than US$1.90 per day. However, mobile phone subscriptions are high (85.2 per cent in 2017, with an increase of 39.4 per cent between 2010 and 2016), providing greater connection within India and to the world. Conversely, internet users comprise a much smaller percentage of the population (only 29.5 per cent in 2017). Despite this, ICT is a boom industry in places like Bengaluru and Hyderabad, where many international companies have set up their service industries providing the world with call centres, and conducting research and development within the ICT sector.

10.6.2 ICT in India

Among Asian countries, India is a leader in internet affordability and is ranked third in its readiness for the internet, but poor mobile speed and uneven availability mean that a digital divide does exist within the country. **FIGURE 1** shows the uneven average download speeds across India. The ICT hubs are within the highest-rated areas, although this rate of connection is lower than can be expected in Melbourne, where the average download speed is over 40 Mbps.

10.6.3 Bengaluru — a dynamic city

Bengaluru began its role in the ICT world back in the 1980s when two Indian tech companies — Infosys and Wipro — moved their head offices there. Other tech companies followed, growing their businesses around the two firms. This included foreign companies looking to cut costs by employing cheap local ICT developers. The ICT outsourcing model had begun.

Bengaluru is now a modern city. These new jobs raised living standards and attracted educated Indians from across the country, as well as expatriates from across the world. Academic institutions set up alongside the innovative ICT businesses. Indians working elsewhere in the world are bringing their knowledge and skills home. More and more international companies are outsourcing to India because labour costs are lower and skill shortages occur across the world. India also has a large and able English-speaking workforce (there are more than 80 million English-speakers in India). In 2019, Australia's Telstra launched its Telstra Innovation and Capability Centre in Bengaluru to overcome the skill shortage in Australia. Bengaluru has grown into a major international hub for ICT companies. Since 2018, Bengaluru and Hyderabad (part of India's Silicon Valley) have shared top billing as the world's most dynamic cities, according to a ranking devised by the investment management firm Jones Lang LaSalle.

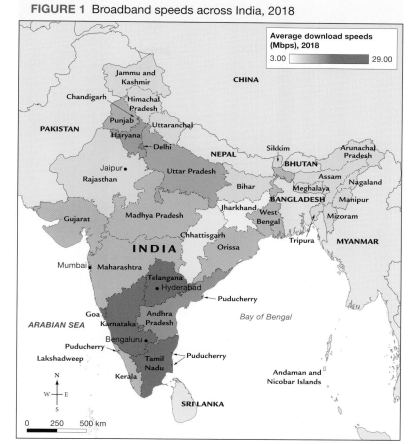

FIGURE 1 Broadband speeds across India, 2018

Average download speeds (Mbps), 2018
3.00 — 29.00

Source: © 2006–2019 Ookla, LLC

FIGURE 2 Modern Bengaluru

FIGURE 3 The strength of India's IT sector

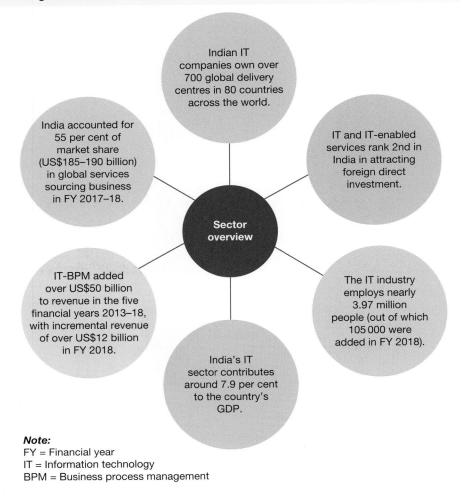

Indian IT companies own over 700 global delivery centres in 80 countries across the world.

IT and IT-enabled services rank 2nd in India in attracting foreign direct investment.

India accounted for 55 per cent of market share (US$185–190 billion) in global services sourcing business in FY 2017–18.

Sector overview

The IT industry employs nearly 3.97 million people (out of which 105 000 were added in FY 2018).

IT-BPM added over US$50 billion to revenue in the five financial years 2013–18, with incremental revenue of over US$12 billion in FY 2018.

India's IT sector contributes around 7.9 per cent to the country's GDP.

Note:
FY = Financial year
IT = Information technology
BPM = Business process management

10.6 EXERCISES

Geographical skills key: GS1 Remembering and understanding **GS2** Describing and explaining **GS3** Comparing and contrasting **GS4** Classifying, organising, constructing **GS5** Examining, analysing, interpreting **GS6** Evaluating, predicting, proposing

10.6 Exercise 1: Check your understanding

1. **GS1** What percentage of the Indian population earns less than US$1.90 per day?
2. **GS1** How rapid was the uptake of mobile phones in India between 2010 and 2016?
3. **GS1** In which Indian cities is the IT industry developing rapidly?
4. **GS1** In what aspects of the IT industry is India particularly well regarded?
5. **GS2** How significant is the ability of the Indian population to speak English? Explain.

10.6 Exercise 2: Apply your understanding

1. **GS2** Look at **FIGURE 1**. Describe the broadband speeds across India.
2. **GS6** Suggest what impact India's broadband speeds would have on the establishment of technological companies across the country.
3. **GS5** In what ways does the ICT sector help the economic development of India within the country?
4. **GS5** In what ways does the ICT sector help the economic development of India with its connections to the world?
5. **GS6** List the advantages of Bengaluru and India to the world as a major ICT hub.

Try these questions in learnON for instant, corrective feedback. Go to www.jacplus.com.au.

10.7 The impact of ICT production

10.7.1 Production and consumption

China is one of the largest producers and consumers of electronics. With the short lifespan of some products — the Chinese buy a new mobile phone on average every 18 months — and with advances in technology, there is a growing amount of e-waste, produced both within China and by overseas countries (**FIGURE 1**). Globally 44.7 million metric tonnes of e-waste were produced in 2016; it is expected this figure will reach 63.7 million metric tonnes by 2025. For a long time, places like China, India and Ghana have accepted and processed the world's e-waste to enhance their economic development.

FIGURE 1 Countries generating the most electronic waste, 2016

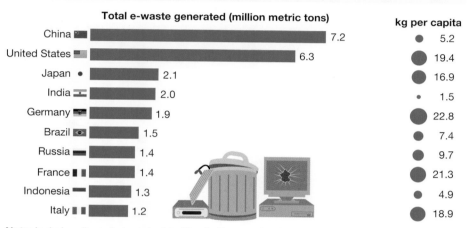

Total e-waste generated (million metric tons)

Country		kg per capita
China	7.2	5.2
United States	6.3	19.4
Japan	2.1	16.9
India	2.0	1.5
Germany	1.9	22.8
Brazil	1.5	7.4
Russia	1.4	9.7
France	1.4	21.3
Indonesia	1.3	4.9
Italy	1.2	18.9

Note: Includes discarded products with a battery or plug including mobile phones, laptops, televisions, refrigerators, electrical toys and other electronic equipment

Source: The Global E-waste Monitor 2017, © Statista

10.7.2 The impact of e-waste on people in China

Growth in China's national economy has seen a change in the sale of ICT appliances as its society develops a growing middle class. China generates the highest quantity of e-waste in Asia and in the world — some 7.2 million metric tonnes in 2016 alone. **FIGURES 2** and **3(a)** and **(b)** show the recent changes in ICT device ownership and disposal of devices in China.

In the domestic market, informal collectors travel door-to-door collecting no-longer-used technological appliances for cash. It is estimated that this mode of collection recovers most e-waste (86 per cent in 2015). Formal collectors are tax-paying businesses or waste stations that buy back old appliances. But the Chinese consumers prefer the informal collectors who offer a higher price and a more convenient service.

The informal collectors' method of handling the e-waste is a major concern for their wellbeing. In backyards and laneways families sift through the e-waste, exposing themselves to many toxic components. **FIGURE 4** shows the various human body systems and the e-waste components that can affect them. Major exposure to the toxic elements occurs when the e-waste component parts are melted down over open fires to extract gold, copper and silver (**FIGURE 5**). Recent studies have shown that exposure to such toxic components reduces intelligence and has a negative impact on the development of the central nervous system of children.

FIGURE 2 The number of ICT devices owned in China

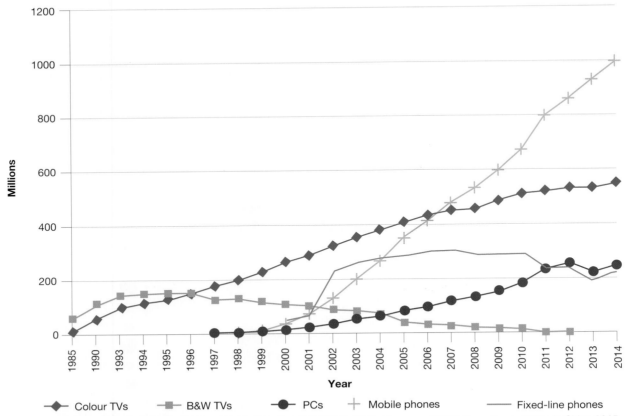

Source: China Household Electric Appliance Research Institute (CHEARI), White Paper on WEEE Recycling Industry in China 2015

FIGURE 3 The number of devices discarded annually in China (a) televisions and PCs and (b) mobile phones

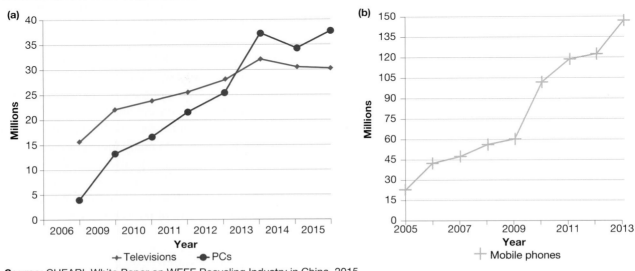

Source: CHEARI, White Paper on WEEE Recycling Industry in China, 2015

FIGURE 4 Health impacts of e-waste on waste workers and people who live near landfills or incinerators

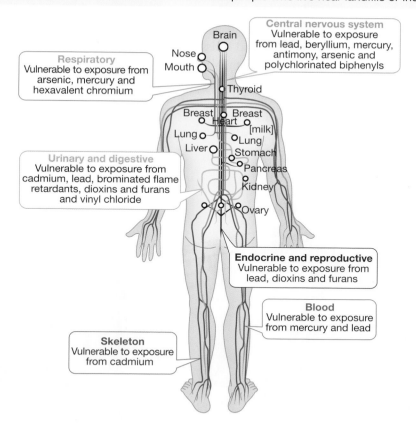

Central nervous system
Vulnerable to exposure from lead, beryllium, mercury, antimony, arsenic and polychlorinated biphenyls

Respiratory
Vulnerable to exposure from arsenic, mercury and hexavalent chromium

Urinary and digestive
Vulnerable to exposure from cadmium, lead, brominated flame retardants, dioxins and furans and vinyl chloride

Endocrine and reproductive
Vulnerable to exposure from lead, dioxins and furans

Blood
Vulnerable to exposure from mercury and lead

Skeleton
Vulnerable to exposure from cadmium

Brain
Nose
Mouth
Thyroid
Breast
Breast
Heart
[milk]
Lung
Lung
Liver
Stomach
Pancreas
Kidney
Ovary

FIGURE 5 Informal collectors sort and burn e-waste.

10.7.3 The impact of e-waste on places in China

For many years Guiyu, in Guangdong province, China, was known as the centre for reclaiming e-waste. The livelihood of its residents depended on this business. The air was polluted by an acidic smell, waste water as a by-product flowed into waterways, and soils were contaminated. Local agricultural produce was contaminated by the toxic water used for irrigation. Vegetables further absorbed toxins through their leaf systems, and people ate these vegetables.

Today Guiyu has a number of modern formal recycling plants. The informal collectors have been forced into operating in and through these plants. However, it has not been easy to change people's ways, so regulation and law enforcement have not always been adequate to bring about change.

FIGURE 6 Animals graze among e-waste in Guiyu.

FIGURE 7 E-waste components for recycling. How we manage our e-waste is an ongoing challenge that will require global solutions.

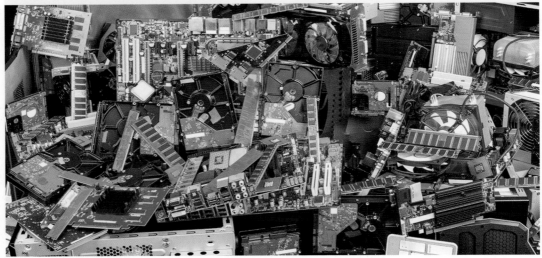

10.7 EXERCISES

Geographical skills key: GS1 Remembering and understanding **GS2** Describing and explaining **GS3** Comparing and contrasting **GS4** Classifying, organising, constructing **GS5** Examining, analysing, interpreting **GS6** Evaluating, predicting, proposing

10.7 Exercise 1: Check your understanding
1. **GS1** Define the term *e-waste*.
2. **GS2** Compile a list of potential e-waste from items in:
    (a) your classroom
    (b) your home.
3. **GS1** Name the two countries that produce the greatest amount of e-waste.
4. **GS1** Name the five countries that produce the most e-waste per person (per capita). How would you categorise the economic development of these countries?
5. **GS2** Describe how the consumption of ICT products has increased in China since 2006.

10.7 Exercise 2: Apply your understanding
1. **GS2** Has the production of e-waste reflected the consumption of ICT products in China since 2006? Explain your answer.
2. **GS6** China might produce the most e-waste, but its per capita level is low. India also has a low per capita level, although it produces far less e-waste. Try to explain this situation.
3. **GS5** Interpret **FIGURE 5**, which shows an informal collector in China sorting and burning e-waste. Annotate the image to show the health effects that this person might experience in the future.
4. **GS5** Explain how the animal grazing in the *environment* in **FIGURE 6** is likely to be affected by the e-waste.
5. **GS6** Propose a set of regulations that might assist the city of Guiyu to replace the culture of informal collection of e-waste in the city. Suggest how each regulation might be introduced so that the program is a success.

Try these questions in learnON for instant, corrective feedback. Go to www.jacplus.com.au.

10.8 The future for e-waste

10.8.1 Addressing concerns

Since 2014, legislation regarding the management of e-waste has been developed and, to varying degrees, adopted across the globe (see **TABLE 1**). The coverage by legislation has risen from 44 per cent to 66 per cent of the world's population (in 67 countries). India, as a major generator of e-waste, has been leading the way with the adoption of legislation; African countries, conversely, have done little to address the issue.

TABLE 1 Global adoption of e-waste legislation

Well-developed e-waste legislation	Absence of e-waste legislation
Europe	Africa
North America	Caribbean
East Asia (China, Hong Kong, Japan, Taiwan, South Korea)	Central Asia
South Asia	East Asia (Mongolia, North Korea)
Oceania (Australia and New Zealand)	Oceania (Pacific Islands)

10.8.2 Legislation

The existence of policies or legislation does not necessarily imply successful enforcement or the existence of sufficient e-waste management systems. **TABLE 2** lists some of the more significant attempts at e-waste management around the world.

TABLE 2 E-waste legislation around the world

Policy/legislation	Specific actions
Basel Convention 1994	• Keep the production of hazardous waste as low as possible. • Make suitable disposal facilities available. • Reduce and manage international flow of hazardous waste. • Ensure management of waste is controlled in an environmentally friendly way. • Block and punish illegal movement of hazardous waste.
Buy-back policies	Many countries have tried buy-back schemes, with varying degrees of success.
China's e-waste ban, 2002	Although an official ban was placed on e-waste being shipped into China, it continued to be smuggled in or came across the borders by land. In 2017 China strengthened its ban on e-waste.
International Telecommunication Union	Connect 2030 has taken on board the Sustainable Development Goals, especially Goals 3, 7, 11, 12 and 13, where ICT can be applied.
Kenya e-waste Act	Initiated in 2013 but stalled in parliament, this Act has been replaced by a National E-Waste Management Strategy to cover the period 2019–20 to 2023–24. Its purpose is to prescribe ways to minimise negative impacts of e-waste on the environment and human health.
Global e-waste Statistics Partnership 2017	The International Telecommunication Union, the United Nations University, and the International Solid Waste Association have joined together to improve the collection, analysis and publication of worldwide e-waste statistics, with a view to increasing the awareness of the need for further development in the e-waste industry.
India 2018	Rules were first established in 2011 using the concept of Extended Producer Responsibility whereby the manufacturer is responsible for safe disposal of electronic goods. In 2018 the emphasis was on regulating the dismantlers and recyclers and providing revised collection targets into the future.

Only 41 countries in the world collect statistics. Measuring e-waste is an important step towards addressing the e-waste challenge. Statistics help to evaluate developments over time, set and assess targets, and identify best practices of policies. Better e-waste data will help to minimise its generation, prevent illegal dumping and emissions, promote recycling and create jobs.

10.8.3 Consumption awareness and responsible e-waste handling

In 2011, the Australian government commenced the National Television and Computer Recycling Scheme (NTCRS). The NTCRS website directs people to places to dispose of e-waste, such as MobileMuster and Planet Ark.

On 1 July 2019 Victoria banned the inclusion of e-waste in general garbage collections and curbside collections. E-waste will no longer go to landfill.

Each individual must be aware of the e-waste being produced by their consumption of modern technological appliances and their method of disposal of no-longer-wanted items. How aware are you? Act local, think global.

FIGURE 1 Toxic components in the central processing unit and cathode-ray tube monitor of a desktop computer

10.8 INQUIRY ACTIVITIES

1. Conduct research to find out where your nearest National Television and Computer Recycling Scheme drop-off point is located.
   (a) Is it realistic to take your e-waste there?
   (b) How can the Victorian government support its legislation on e-waste?

   Examining, analysing, interpreting

2. As a class, conduct a survey and interview students, teachers and families about their e-waste recycling habits.
   (a) Draft the questions you wish to ask and consider how you will record the responses. Some ideas for questions might include whether students are more **environmentally** aware than teachers/families, whether age makes a difference to a person's attitude to e-waste recycling, whether you consider you have enough e-waste to make it worthwhile recycling, whether it is easy to reach a location that will take your e-waste, etc.
   (b) If you wish to conduct your survey online, use the **Survey Monkey** weblink in the Resources tab. Otherwise, you can use the SkillBuilder in subtopic 10.10 to assist in your survey development.
   (c) After you have conducted your surveys, collate and present your findings in graphic form.
   (d) Analyse the graphs and write a summary of the findings. If possible, arrange to present your findings to an interested group within your school or community — you may help improve awareness of issues and **change** attitudes towards e-waste recycling!

   Classifying, organising, constructing

3. Research and write a considered paragraph on the state of e-waste management in the United States, Germany, Thailand or Nigeria.

   Examining, analysing, interpreting

10.8 EXERCISES

Geographical skills key: GS1 Remembering and understanding **GS2** Describing and explaining **GS3** Comparing and contrasting **GS4** Classifying, organising, constructing **GS5** Examining, analysing, interpreting **GS6** Evaluating, predicting, proposing

10.8 Exercise 1: Check your understanding

1. **GS1** What proportion of the world's countries has legislation in place regarding e-waste management?
2. **GS1** Which areas of the world are lacking in e-waste management legislation?
3. **GS1** Outline the key actions identified in the Basel Convention.
4. **GS2** Explain the importance of statistics in addressing the issue of e-waste management.
5. **GS1** List five toxic components within a desktop computer.

10.8 Exercise 2: Apply your understanding

1. **GS6** Why does legislation often seem to have limited impact in the e-waste sector?
2. **GS6** Now more than 25 years on from the Basel Convention, how has the world responded to the legislation?
3. **GS6** Suggest reasons for Kenya's inability to bring into law an e-waste Act.
4. **GS6** Many countries are looking at an Extended Producer Responsibility (EPR) for e-waste. In China, the four key areas of manufacturing responsibility are: producing *environmentally* friendly designs; using recycled materials; standardising waste management and recycling processes; and disclosing data on recycling. Discuss whether these four aspects of e-waste management are likely to be easily, readily and willingly taken into law in China.
5. **GS5** Explain what is meant by the 'need for a global solution to the transboundary issue of e-waste'.

Try these questions in learnON for instant, corrective feedback. Go to www.jacplus.com.au.

10.9 SkillBuilder: Constructing a table of data for a GIS

Why are there tables within GIS?

Geographical information systems, or GIS, use tables to organise and store information about points, lines, and polygons (vector data). These tables have rows and columns, called fields. The GIS software links the rows in the table to the points, lines or polygons on a map.

Select your learnON format to access:

- an overview of the skill and its application in Geography (Tell me)
- a video and a step-by-step process to explain the skill (Show me)
- an activity and interactivity for you to practise the skill (Let me do it)
- questions to consolidate your understanding of the skill.

Sample	Address	No_home	No_mobiles
1	42 Jacob Street	2	4
2	27 Jacob Street	3	3
3	36 Adele Avenue	4	3
4	34 Flint Street	4	1
5	35 Flint Street	5	3
6	25 Flint Street	4	2
7	12 Jess Court	4	2
8	2 Jess Court	4	4
9	12 Flint Street	5	3
10	52 Jacob Street	6	2

On Resources

Video eLesson Constructing a table of data for a GIS (eles-1743)

Interactivity Constructing a table of data for a GIS (int-3361)

10.10 SkillBuilder: Using advanced survey techniques — interviews

What are interviews that survey people's opinions?

Surveys collect primary data, such as data that has been gathered in the field. Conducting a survey interview means asking questions, recording and collecting responses, and collating the number of responses.

Select your learnON format to access:

- an overview of the skill and its application in Geography (Tell me)
- a video and a step-by-step process to explain the skill (Show me)
- an activity and interactivity for you to practise the skill (Let me do it)
- questions to consolidate your understanding of the skill.

 Resources

Video eLesson Using advanced survey techniques — interviews (eles-1742)

Interactivity Using advanced survey techniques — interviews (int-3360)

10.11 Thinking Big research project: Trash or treasure?

SCENARIO

Showcasing Japanese dedication to sustainability, the Tokyo 2020 Olympic medals contain electronic waste. You will create a pamphlet to accompany the medals, explaining the background to their production — how the trash of millions has been recycled to create the prized Olympic treasures of the athletes of the 2020 Olympic Games.

Select your learnON format to access:

- the full project scenario
- details of the project task
- resources to guide your project work
- an assessment rubric.

Resources

ProjectsPLUS Thinking Big research project: Trash or treasure? (pro-0196)

10.12 Review

10.12.1 Key knowledge summary
Use this dot point summary to review the content covered in this topic.

10.12.2 Reflection
Reflect on your learning using the activities and resources provided.

Resources

eWorkbook   Reflection (doc-31730)

Crossword (doc-31731)

Interactivity   Global ICT — connections, disparity and impacts crossword (int-7652)

KEY TERMS

connectivity the ability to access the internet

digital divide a type of inequality between groups in their access to and knowledge of information and communication technology

e-waste any old electrical equipment such as computers, toasters, mobile phones and iPods that no longer works or is no longer required

gross domestic product (GDP) the value of all the goods and services produced within a country in a given period of time (usually a year). It is often used as an indicator of a country's wealth.

human development measures such as life expectancy, education and economic wellbeing that provide an overall indication of a place's level of development and the standard of living of its inhabitants

World Wide Web the global resources and information exchange available to internet users through the use of the Hypertext Transfer Protocol (HTTP)

GLOSSARY

active travel making journeys via physically active means such as cycling or walking

agribusiness business set up to support, process and distribute agricultural products

agroforestry the use of trees and shrubs on farms for profit or conservation; the management of trees for forest products

anthropogenic resulting from human activity (man-made)

aquaculture the farming of aquatic plants and aquatic animals such as fish, crustaceans and molluscs

aquaponics a sustainable food production system in which waste produced by fish or other aquatic animals supplies the nutrients for plants, which in turn purify the water

aquifer a body of permeable rock below the Earth's surface, which contains water, known as groundwater

arable describes land that can be used for growing crops

barter to trade goods in return for other goods or services rather than money

biodiversity the variety of plant and animal life within an area

biofuel fuel that comes from renewable sources

biophysical environment the natural environment, made up of the Earth's four spheres — the atmosphere, biosphere, lithosphere and hydrosphere

black market any illegal trade in officially controlled or scarce goods

canal housing estate a housing estate built upon a system of waterways, often as the result of draining wetland areas. All properties have water access.

cash crop a crop grown to be sold so that a profit can be made, as opposed to a subsistence crop, which is for the farmer's own consumption

clear-felling the removal of all trees in an area

connectivity the ability to access the internet

coral polyp a tube-shaped marine animal that lives in a colony and produces a stony skeleton. Polyps are the living part of a coral reef.

crop rotation a procedure that involves the rotation of crops, so that no bed or plot sees the same crop in successive seasons

deforestation clearing forests to make way for housing or agricultural development

degradation deterioration in the quality of land and water resources caused by excessive exploitation

desertification the transformation of arable land into desert, which can result from climate change or from human practices such as deforestation and overgrazing

developed describes countries with a highly developed industrial sector, a high standard of living, and a large proportion of people living in urban areas

developing countries nations with a low living standard, undeveloped industrial base and low human development index relative to other countries

digital divide a type of inequality between groups in their access to and knowledge of information and communication technology

disability a functional limitation in an individual, caused by physical, mental or sensory impairment

discretionary item an item that is bought out of choice, according to one's judgement

e-waste any old electrical equipment such as computers, toasters, mobile phones and iPods that no longer works or is no longer required

ecotourism tourism that interprets the natural and cultural environment for visitors, and manages the environment in a way that is ecologically sustainable

endemic describes species that occur naturally in only one region

environmental refugees people who are forced to flee their home region due to environmental changes (such as drought, desertification, sea-level rise or monsoons) that affect their wellbeing or livelihood

erosion the wearing down of rocks and soils on the Earth's surface by the action of water, ice, wind, waves, glaciers and other processes

ethnicity cultural factors such as nationality, culture, ancestry, language and beliefs

extensive farm farm that extends over a large area and requires only small inputs of labour, capital, fertiliser and pesticides

extremism extreme political or religious views or extreme actions taken on the basis of those views

factory farming the raising of livestock in confinement, in large numbers, for profit

genetically modified describes seeds, crops or foods whose DNA has been altered by genetic engineering techniques

graffiti the marking of another person's property without permission; it can include tags, stencils and murals

Green Revolution a significant increase in agricultural productivity resulting from the introduction of high-yield varieties of grains, the use of pesticides and improved management practices

greenhouse gases any of the gases that absorb solar radiation and are responsible for the greenhouse effect. These include water vapour, carbon dioxide, methane, nitrous oxide and various fluorinated gases.

gross domestic product (GDP) the value of all the goods and services produced within a country in a given period of time (usually a year). It is often used as an indicator of a country's wealth.

groundwater water that exists in pores and spaces in the Earth's rock layers, usually from rainfall slowly filtering through over a long period of time

halal describes food that is prepared under Islamic dietary guidelines

hard sport tourism tourism in which someone travels to either actively participate in or watch a competitive sport as the main reason for their travel

horticulture the practice of growing fruit and vegetables

human development measures such as life expectancy, education and economic wellbeing that provide an overall indication of a place's level of development and the standard of living of its inhabitants

humanitarian principles the principles governing our response to those in need, with the main aim being to save lives and alleviate suffering

humus an organic substance in the soil that is formed by the decomposition of leaves and other plant and animal material

hybrid plant or animal bred from two or more different species, sub-species, breeds or varieties, usually to attain the best features of the different stocks

hydroponic describes a method of growing plants using mineral nutrients, in water, without soil

income diversity income that comes from many sources

indicator something that provides a pointer, especially to a trend

infrastructure the facilities, services and installations needed for a society to function, such as transportation and communications systems, water pipes and power lines

innovation new and original improvement to something, such as a piece of technology or a variety of plant or seed

intensive farm farm that requires a lot of inputs, such as labour, capital, fertiliser and pesticides

irrigation the supply of water by artificial means to agricultural areas

jatropha any plant of the genus *Jatropha*, but especially *Jatropha curcas*, which is used as a biofuel

kenaf a plant in the hibiscus family that has long fibres; useful for making paper, rope and coarse cloth

land degradation deterioration in the quality of land resources caused by excessive exploitation

latitude the angular distance north or south from the equator of a point on the Earth's surface

leaching the process by which water runs through soil, dissolving minerals and carrying them into the subsoil

leeward describes the area behind a mountain range, away from the moist prevailing winds

logging large-scale cutting down, processing and removal of trees from an area

mallee vegetation areas characterised by small, multi-trunked eucalypts found in the semi-arid areas of southern Australia

malnourished describes someone who is not getting the right amount of the vitamins, minerals and other nutrients to maintain healthy tissues and organ function

marginal land describes agricultural land that is on the margin of cultivated zones and is at the lower limits of being arable

Masai an ethnic group of semi-nomadic people living in Kenya and Tanzania

mature-aged describes individuals aged over 55

median age the age that is in the middle of a population's age range, dividing a population into two numerically equal groups

mobility the ability to move or be moved freely and easily

monoculture the cultivation of a single crop on a farm or in a region or country

national park a park or reserve set aside for conservation purposes

national security the protection of a nation's citizens, natural resources, economy, money, environment, military, government and energy

new moon the phase of the moon when it is closest to the sun and is not normally visible

non-government organisation (NGO) a group or business that is organised to serve a particular social purpose at local, national or international level, and operates independently of government

offshore to relocate part of a company's processes or services overseas in order to decrease costs

old-growth forests natural forests that have developed over a long period of time, generally at least 120 years, and have had minimal unnatural disturbance such as logging or clearing

organic matter decomposing remains of plant or animal matter

per capita per person

perception the process by which people translate sensory input into a view of the world around them

plantation an area in which trees or other large crops have been planted for commercial purposes

pneumatophores exposed root system of mangroves, which enables them to take in air when the tide is in

potable drinkable; safe to drink

prairie native grassland of North America

precipitation the forms in which moisture is returned to the Earth from the sky, most commonly in the form of rain, hail, sleet and snow

pulp the fibrous material extracted from wood or other plant material to be used for making paper

rain shadow the dry area on the leeward side of a mountain range

Ramsar site a wetland of international importance, as defined by the Ramsar Convention — an intergovernmental treaty on the protection and sustainable use of wetlands

salinity the presence of salt on the surface of the land, in soil or rocks, or dissolved in rivers and groundwater

smuggling importing or exporting goods secretly or illegally

social justice a principle applied so that a society is based on equality, the appreciation of the value of human rights and the recognition of the dignity of every human being

soft sport tourism tourism in which someone participates in recreational and leisure activities, such as skiing, fishing and hiking as part of their travel

stereotype widely held but oversimplified idea of a type of person or thing

street art artistic work done with permission from both the person who owns the property on which the work is being done and the local council

sustainable describes the use by people of the Earth's environmental resources at a rate such that the capacity for renewal is ensured

totem an animal, plant, landscape feature or weather pattern that identifies an individual's connection to the land

trade barrier government-imposed restriction (in the form of tariffs, quotas and subsidies) on the free international exchange of goods or services

trading partner a participant, organisation or government body in a continuing trade relationship

treaty a formal agreement between two or more independent states or nations, and usually involving a signed document

treeline the edge of the area in which trees are able to grow

tundra the area lying beyond the treeline in polar or alpine regions

turn-up-and-go frequent and regular transport service such that reference to a timetable is not required; e.g. users know that a train will run every 10 minutes

undernourished describes someone who is not getting enough calories in their diet; that is, not enough to eat

undulating describes an area with gentle hills

urban agglomeration the extended built-up area of a place, including suburbs and continuous urban area

urbanisation the growth and spread of cities

value adding processing a material or product and thereby increasing its market value

water stress situation that occurs when water demand exceeds the amount available or when poor quality restricts its use

waterlogging saturation of the soil with groundwater such that it hinders plant growth

watertable the surface of the groundwater, below which all pores in the soils and rock layers are saturated with water

Western-style diet eating pattern common in developed countries, with high amounts of red meat, sugar, high-fat foods, refined grains, dairy products, high-sugar drinks and processed foods

windward describes the side of the mountain that faces the prevailing winds

winter solstice the shortest day of the year, when the sun reaches its lowest point in relation to the equator

World Wide Web the global resources and information exchange available to internet users through the use of the Hypertext Transfer Protocol (HTTP)

yield gap the gap between a certain crop's average yield and its maximum potential yield

INDEX

Note: Figures and tables are indicated by italic *f* and *t*, respectively, following the page reference.